HANDBOOK OF SOVIET LUNAR AND PLANETARY EXPLORATION

PUBLICATIONS OF THE AMERICAN ASTRONAUTICAL SOCIETY

Following are the principal publications of the American Astronautical Society:

JOURNAL OF THE ASTRONAUTICAL SCIENCES (1954-)

Published quarterly and distributed by AAS Business Office, 6060 Duke Street, Alexandria, VA 22304

AAS NEWSLETTER (1962-)

Published bimonthly and distributed by AAS Business Office, 6060 Duke Street, Alexandria, VA 22304

ADVANCES IN THE ASTRONAUTICAL SCIENCES (1957-)

Proceedings of major AAS technical meetings. Published and distributed for the American Astronautical Society by Univelt, Inc., P.O. Box 28130, San Diego, CA 92128

SCIENCE AND TECHNOLOGY SERIES (1964-)

Supplement to Advances in the Astronautical Sciences. Proceedings and monographs, most of them based on AAS technical meetings. Published and distributed for the American Astronautical Society by Univelt, Inc., P.O. Box 28130, San Diego, CA 92128

AAS HISTORY SERIES (1977-)

Supplement to Advances in the Astronautical Sciences. Selected works in the field of aerospace history under the editorship of Dr. Eugene M. Emme, NASA Historian. Published and distributed for the American Astronautical Society by Univelt, Inc., P.O. Box 28130, San Diego, CA 92128

AAS MICROFICHE SERIES (1968-)

Supplement to Advances in the Astronautical Sciences. Consists principally of technical papers not included in the hard-copy volumes. Published and distributed for the American Astronautical Society by Univelt, Inc., P.O. Box 28130, San Diego, CA 92128

Subscriptions to the Journal and the AAS Newsletter should be ordered from the AAS Business Office. Back issues of the Journal and all books and microfiche should be ordered from Univelt, Inc.

Front Cover Illustration:

Lunokhod 1 shown on top of the Luna 17 descent stage ready for descent to lunar surface (R. Escarcega)
See Figure 35, page 72.

Frontispiece:

Luna 16's retrieval of a sample of lunar soil and blast-off from the moon (R. Escarcega)
See Figure 29, page 60.

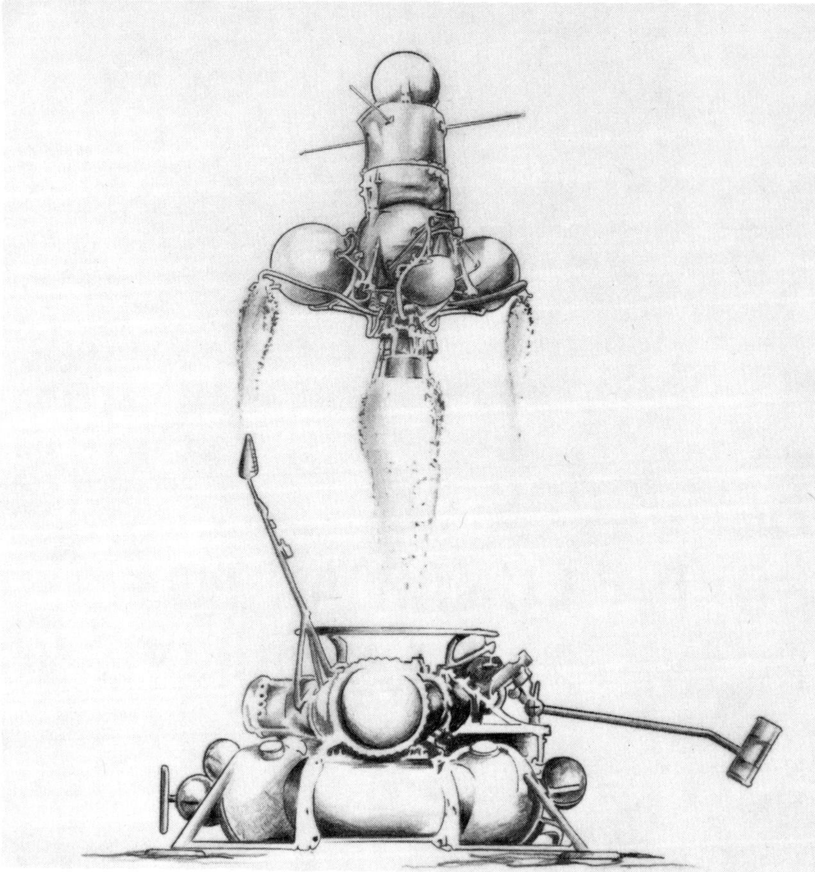

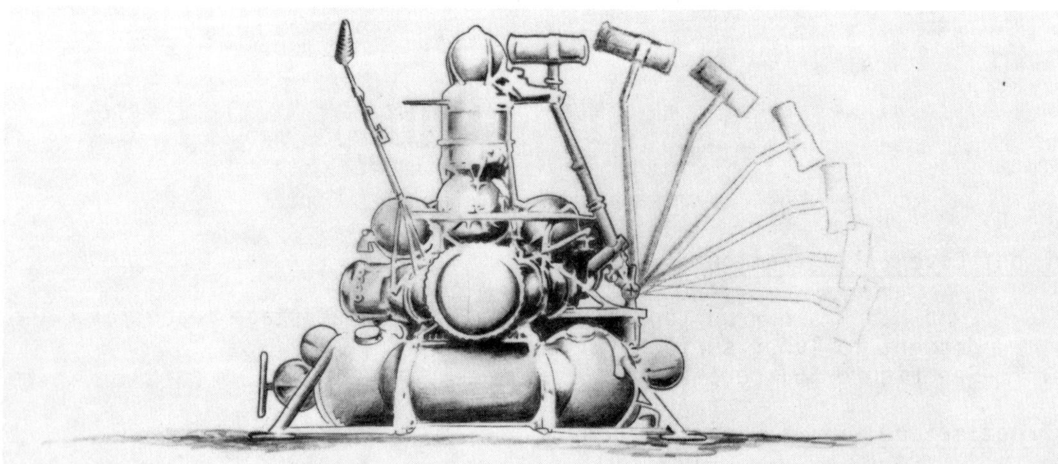

AN AMERICAN ASTRONAUTICAL SOCIETY PUBLICATION

HANDBOOK OF SOVIET LUNAR AND PLANETARY EXPLORATION

Nicholas L. Johnson

Volume 47
SCIENCE AND TECHNOLOGY SERIES

A Supplement to Advances in the Astronautical Sciences

1979

Published by Univelt, Inc., P.O. Box 28130
San Diego, California 92128

Affiliated with the American Association for the Advancement of Science
Member of the International Astronautical Federation

ISBN 0-87703-105-3 (Hard Cover)
ISBN 0-87703-106-1 (Soft Cover)

Published for the American Astronautical Society by
Univelt, Inc., P. O. Box 28130, San Diego, California 92128

Printed and Bound in the U.S.A.

FOREWORD

"Mankind will not forever remain on earth, but in the pursuit of light and space will first timidly emerge from the bounds of the atmosphere, and then advance until he has conquered the whole of circumsolar space."

K. E. Tsiolkovskiy, 1911

Six years before the great October Revolution was to spawn the Union of Soviet Socialist Republics, these words were penned by K. E. Tsiolkovskiy, the founder and prophet of Soviet astronautics. Although the principles of rocket propulsion had been known for hundreds of years, not until the 1903 publication of Tsiolkovskiy's *Investigation of Cosmic Spaces by Reactive Devices* was there a scientific analysis of the obstacles to be overcome if man were to leave his island in space. Fifty-four years later, relying on the groundwork set forth by Tsiolkovskiy, the Soviet Union launched another October revolution, the space revolution. On October 4, 1957, the beep-beep-beep of Sputnik 1 thrust a reluctant world into the Space Age.

Like a modern day Columbus, Yuri Gagarin fulfilled Tsiolkovskiy's prophecy a scant three and one-half years later by becoming the first of mankind to "emerge from the bounds of the atmosphere" into earth orbit. But there was nothing timid in the Soviet Union's drive to conquer space. Since the launching of Sputnik 1 in 1957 one space spectacular has followed another in an astounding Soviet effort designed to reach that stellar destiny of the Soviet people.

The first earth satellite, the first manned spacecraft, the first lunar and planetary probes all carried the familiar hammer and sickle emblem. Soviet cosmonauts were the first to walk in space, conduct crew transfers between spaceships, and operate scientific space stations for extended periods. The first spacecraft to land on the moon, Mars, and Venus were built by Soviet craftsman. Although no Soviet cosmonauts have yet traversed the lunar surface, automatic Soviet robots have roved the lunar seas for periods exceeding one year and have returned soil samples from topographically diverse regions.

It was, of course, the unexpectedness of the Soviet Union's sudden prowess in space exploration which sent the rest of the world reeling. Almost a decade passed before the United States could recover from this cosmic Pearl Harbor and gain that elusive and ill-defined "lead" in space. Although the space race officially ended on July 20, 1969, when an American flag, not a Soviet banner, was erected by the first men on the moon, the competition in outer space continues today. Manned Soviet space stations now orbit overhead while the United States prepares to enter a new age of exploration with the reusable Space Shuttle.

What then may be the future role of the Soviet Union in space? Will the Soviets continue their broad space programs at presently high levels while the United States is reducing her own involvement in space exploration? Unfortunately, the answers to these questions must be deferred until such a time as the Soviet Union feels that the need for the secrecy surrounding her space activities has diminished. However, to appreciate the motives behind the Soviet Union's relentless drive to the moon and planets, and perhaps to catch a glimpse of what may lie in the future, one must first analyze her past accomplishments and goals in the proper perspective.

Since the dawn of the Space Age over twenty-two years ago hundreds of volumes have chronicled the United States' efforts to reach the moon and beyond. Whereas there have been also numerous attempts to construct a similar history of Soviet space exploration, no one source has yet examined in depth the Soviet programs for manned and planetary exploration. This volume has been written to fill that void for the lunar and planetary robot explorers of the USSR. A companion volume will be devoted to the history of Soviet manned space flight.

Any work purporting to portray accurately the evolution of Soviet space exploration is immediately beset by a question of credibility. Although Soviet space scientists and journalists have the advantage of first-hand knowledge of Soviet intentions and goals, seldom are these reflected in Soviet publications or at international conferences. While technical information pertaining to spacecraft and other hardware has been found with few exceptions to be accurate, the same cannot be said of Soviet announced programs prior to or after their implementation. Here the record is littered by misleading and sometimes false statements. Even more glaring in Soviet authored accounts are the omissions of fact and purpose.

On the other hand, many Western documents concerned with the Soviet space program are largely a reflection of Soviet propaganda and of their author's beliefs and deductions which may or may not adhere to reality. In addition, the purpose of many of these publications is to condemn or praise the achievements of the USSR rather than to report them objectively. This serves merely to confuse further the true picture. However, scholarly studies are conducted in the West by individuals and groups, and these studies have proved invaluable to Soviet space watchers.

This then is the basis from which every author must begin. To ensure the most complete and reliable investigation thousands of books, pamphlets, articles, and documents have been examined. Whenever possible Russian language publications were obtained through the National Technical Information Service of the US Department of Commerce for historical purposes and to verify scientific data. Domestic and foreign periodicals and English-language Soviet publications also were reviewed for informative articles by Soviet space scientists. Finally, a wide range of Western works was consulted and personal correspondence was conducted with acknowledged experts in the field.

In particular, mention must be made of the efforts and publications of the British Interplanetary Society, who in addition to its prominent role in the support of space exploration is the acknowledged leader of Soviet space-flight analysis.

Great care has been taken to present in this volume only the facts as they now appear. When minor inconsistencies in data due to the continuing refinement of telemetry occurred, the most recent and widely accepted figures have been incorporated. Rarely are conjectures brought forth, but when appropriate they are always labeled as such. To this end the application of footnotes and references has been widely adopted. The cited reference is important not only to judge its own reliability, but also as a source for further information on the topic in question.

This work consists of two parts: Volumes 47 and 48, AAS Science and Technology Series. Volume 47, titled *Handbook of Soviet Lunar and Planetary Exploration,* deals essentially with that portion of the Soviet unmanned space program devoted to the exploration of the moon, Venus, and Mars. Volume 48, titled *Handbook of Soviet Manned Space Flight,* presents a look at the Soviet manned space program. Each volume contains pertinent appendices and an index.

The Soviet exploration of the moon and planets by automatic spacecraft is subdivided into four sections: the Luna , Zond, Venera, and Mars programs. Each is covered in depth with emphasis on the increasing sophistication and versatility of each successive probe and the subsequent scientific findings.

The Appendices contain supplementary material which is common to more than one program, or which is summary in nature. In particular, Appendix I and Appendix II, respectively, describe Soviet launch vehicles and launch facilities employed by many programs. Appendix III provides quick reference to all lunar and planetary explorations by the USSR.

Volume 48 examines the evolution of Soviet manned space flight from the brief and heroic flight of Yuri Gagarin aboard Vostok 1 to the latest Soyuz — Salyut marathon space ventures. Each program--Vostok, Voskhod, and Soyuz--is reviewed from the standpoint of spacecraft design, development, and operation. A summary of each flight and related flights is also provided

A work of this magnitude is only possible with the help of many individuals and organizations. Special thanks are due to the staff of the Smithsonian Institution's National Air and Space Museum, the National Aeronautics and Space Administration, and the Foreign Technology Division at Wright-Patterson AFB, Ohio. My personal gratitude is extended to Anthony Kenden and Phillip S. Clark, whose criticisms and contributions have been invaluable, and to Ralph Gibbons and Richard Escarcega for their outstanding artwork.

But most of all I wish to thank my wife, Beth, for her enduring love and support without which these volumes could never have been written.

Nicholas L. Johnson

CONTENTS

CONTENTS (Cont'd)

CONTENTS (Cont'd)

CONTENTS (Cont'd)

Section I
EXPLORATION OF THE MOON
LUNA SERIES

INTRODUCTION

During the over twenty years of Soviet lunar exploration, the USSR has achieved almost every significant "first" with the exception of a manned lunar landing:

1. First spacecraft to fly by the moon (Luna 1)
2. First spacecraft to hit the moon (Luna 2)
3. First spacecraft to photograph the far side of the moon (Luna 3)
4. First spacecraft to soft-land on the moon (Luna 9)
5. First lunar satellite (Luna 10)
6. First chemical and mechanical analysis of the lunar soil (Luna 13)
7. First automated return of lunar soil (Luna 16)
8. First mobile lunar explorer (Luna 17).

As with any endeavor on the forefront of technolgoy, these spectacular successes came only as the result of an enormous amount of collective effort and personal dedication. Unlike the areas of manned and planetary exploration in which each successive flight was dependent upon its predecessors, the lunar program appears to have been one of isolated flights with great successes and great failures. The flights themselves seem to have come in spurts and stops. In the over twenty years since Luna 1, almost all the significant achievements have occurred in just three separate years: Lunas 1-3, 1959; Lunas 9-13, 1966; Lunas 16-17, 1970.

The Soviet Union conducted two simultaneous unmanned lunar exploration programs. The first, the Luna series, which has run from 1959 to the present, is aimed at uncovering the physical nature of the moon and its surrounding space both for scientific knowledge and to aid in the development of a manned lunar program. The second series, the man-oriented Zond series, which lasted from 1968 to 1970 (Zond 1-3 not included), was designed to perfect the actual hardware and spacecraft needed for a manned lunar landing. This section deals solely with the efforts of the Luna series. The Zond program is described in the next section.

LUNAR PROBE SUMMARY

Designator	Date Launched	Mission	Total Weight	Probe Weight	Results
Luna 1	2 Jan 1959	Flyby?	361.3 kg	361.3 kg	Passed moon at 5000-6000 km on 4 Jan
Luna 2	12 Sep 1959	Impact	390.2	390.2	First lunar impact on 14 Sep
Luna 3	4 Oct 1959	Far side pictures	435	278.5	Photographed far side on 7 Oct
Unannounced	4 Jan 1963	Soft land?	1420?	100?	Failed to leave earth orbit
Luna 4	2 Apr 1963	Soft-land	1422	100?	Passed moon at 8500 km on 6 Apr
Kosmos 60	12 Mar 1965	Soft-land	1475?	100?	Failed to leave earth orbit
Luna 5	9 May 1965	Soft-land	1476	100	Crashed on moon on 12 May
Luna 6	8 Jun 1965	Soft-land	1442	100	Passed moon at 161,000 km on 11 Jun
Luna 7	4 Oct 1965	Soft-land	1506	100	Crashed on moon on 8 Oct
Luna 8	3 Dec 1965	Soft-land	1552	100	Crashed on moon on 7 Dec
Luna 9	31 Jan 1966	Soft-land	1583	100	First lunar soft-landing on 3 Feb; returned pictures
Kosmos 111	1 Mar 1966	?	?	?	Failed to leave earth orbit
Luna 10	31 Mar 1966	Lunar orbit	1600	245	First lunar satellite on 3 Apr
Luna 11	24 Aug 1966	Lunar orbit	1640	1135?	Entered lunar orbit on 28 Aug
Luna 12	22 Oct 1966	Lunar orbit photographer	1660?	1136	Entered lunar orbit on 25 Oct; returned pictures
Luna 13	21 Dec 1966	Soft-land	1600?	109	Soft-landed on 24 Dec; measured chemical properties of soil; returned pictures
Luna 14	7 Apr 1968	Lunar orbit	1660?	1140?	Entered lunar orbit on 10 Apr
Luna 15	13 Jul 1969	Retriever or Rover?	5700?	?	Crashed on moon on 21 Jul
Kosmos 300	23 Sep 1969	Retriever or Rover?	5700?	?	Failed to leave earth orbit
Kosmos 305	22 Oct 1969	Retriever or Rover?	5700?	?	Failed to leave earth orbit
Luna 16	12 Sep 1970	Soil Retriever	5800? kg	1880 kg	Landed on moon on 20 Sep; left moon on 21 Sep; landed on earth on 24 Sep with soil sample
Luna 17	10 Nov 1970	Lunokhod (rover)	5700?	1836 (756)	Landed on moon on 15 Nov; returned data until 4 Oct 1971
Luna 18	2 Sep 1971	Soil Retriever	5900?	1880	Crashed in lunar mountains on 11 Sep
Luna 19	28 Sep 1971	Lunar orbit photographer	5900?	1950?	Entered lunar orbit on 3 Oct; returned pictures
Luna 20	14 Feb 1972	Soil Retriever	5900?	1880	Landed on moon on 21 Feb; left moon on 23 Feb; landed on earth on 25 Feb with soil sample
Luna 21	8 Jan 1973	Lunokhod (rover)	5950?	1880 (840)	Landed on moon on 16 Jan; returned data until 3 Jun 1973
Luna 22	29 May 1974	Lunar orbit photographer	5900?	1950?	Entered lunar orbit around 3 Jun; returned pictures
Luna 23	28 Oct 1974	Soil Retriever	6000?	1950?	Landed on moon on 6 Nov; drilling arm damaged; did not return
Luna 24	9 Aug 1976	Soil Retriever	6000?	1950?	Landed on moon on 18 Aug; left moon on 19 Aug; landed on earth on 23 Aug with soil sample

Figure 1a

Figure 1a lists all the acknowledged and suspected Luna-type spacecraft to achieve at least earth orbit. The flights appearing in Figure 1b are suspected launch failures which did not achieve earth orbit. Since the USSR has only very recently even admitted to launch failures, those listings in Figure 1b are necessarily subject to debate. All supporting data and references are covered in the accompanying text.

Figure 2 is a map of the visible side of the moon indicating the location of each Soviet spacecraft's landing site.

POSSIBLE LUNA SERIES
SPACECRAFT WHICH FAILED DURING LAUNCH

Date Launched	Probable Mission
1 May 1958	Flyby or impact
24 Sep 1958	Flyby or impact
26 Nov 1958	Flyby or impact
Feb 1963	Soft land
Apr 1964	Soft land
Apr 1965	Soft land
Apr 1969	Soil retriever or Lunokhod rover
14 Jun 1969	Soil retriever or Lunokhod rover
19 Feb 1970	Soil retriever or Lunokhod rover

Figure 1b

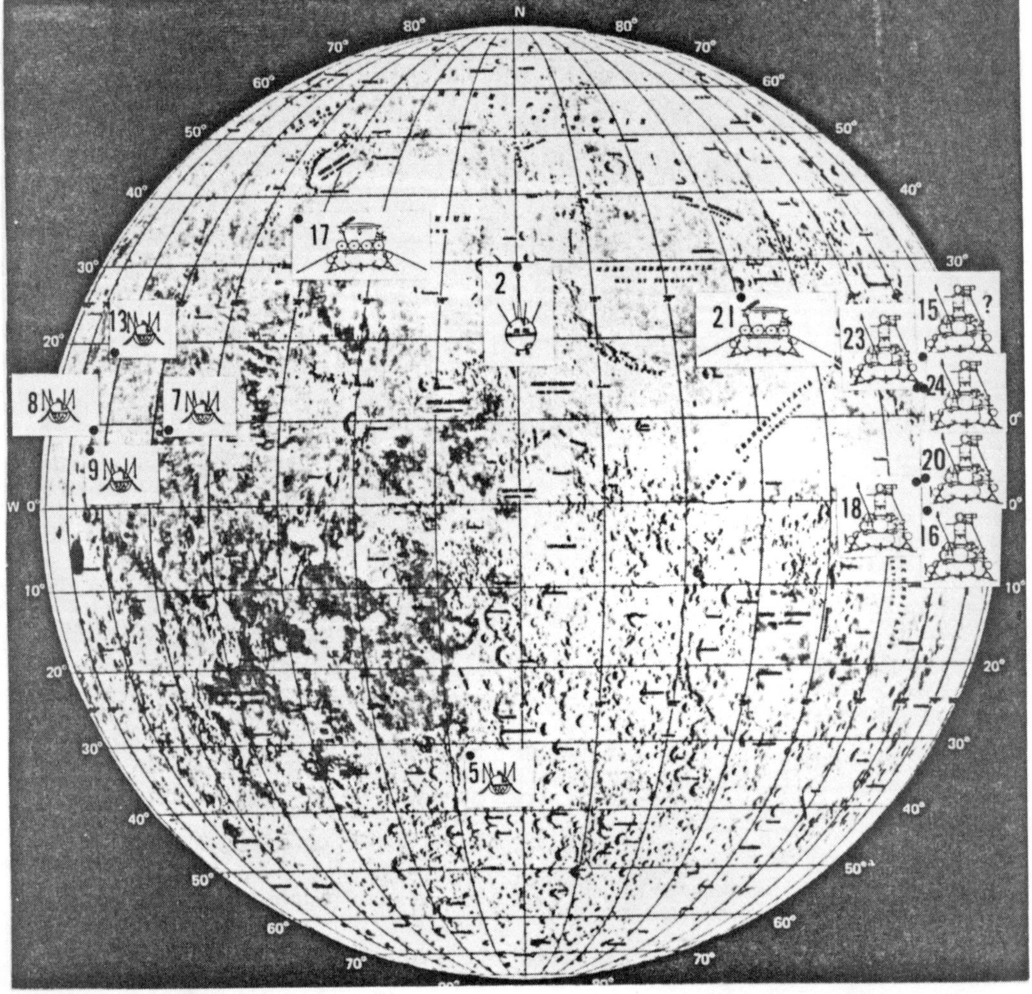

Figure 2 Landing sites of spacecraft of the Luna series

Chapter 1

LUNA SERIES: FIRST-GENERATION SPACECRAFT

LUNA 1

On 2 January 1959 the first man-made object to escape the influence of the earth was launched from deep within the Soviet Union. After witnessing many attempts by the United States end in failure, the world was once again surprised by the boldness of the USSR. After launching Sputnik 3 on 15 May 1958, the USSR's exploration of space had fallen silent.[1]

Luna 1 (also known as Mechta) was inserted directly on a lunar trajectory without first entering an earth orbit (Figure 3). The 120-cm diameter spherical probe closely resembled Sputnik 1 which had ushered in the Space Age a mere fifteen months earlier (Figure 4). Approximately 34 hours after launch the spacecraft passed over the lunar surface at a distance of 5000 - 6000 km and became the first man-made satellite of the sun, entering a 450-day solar orbit (Figure 5). Communications with the probe ceased after 62 hours at a distance of 597,000 km from the earth.[2]

The interior of the Luna 1 capsule, which weighed 361.3 kg, was pressurized to 1.3 atmospheres and was kept at a temperature of about 20°C. The temperature-control system consisted of forced circulation of the interior gas to transfer heat from the scientific instruments to the shell of the probe where it could then be radiated into space. All instruments and radio transmitter-receivers were powered by chemical storage batteries within the probe.

Luna 1 was designed to perform a variety of experiments during its flight. Installed on or within the probe were the following instruments:

1. Apparatus for radio-control of the rocket's flight path, consisting of transmitter, operating on frequency 183.6 Mhz, and receiver block
2. Radio-transmitter, operating on frequency 19.993 Mhz
3. Telemetry block for transmitting research data and data of pressure and temperature within the container to the earth by radio-systems

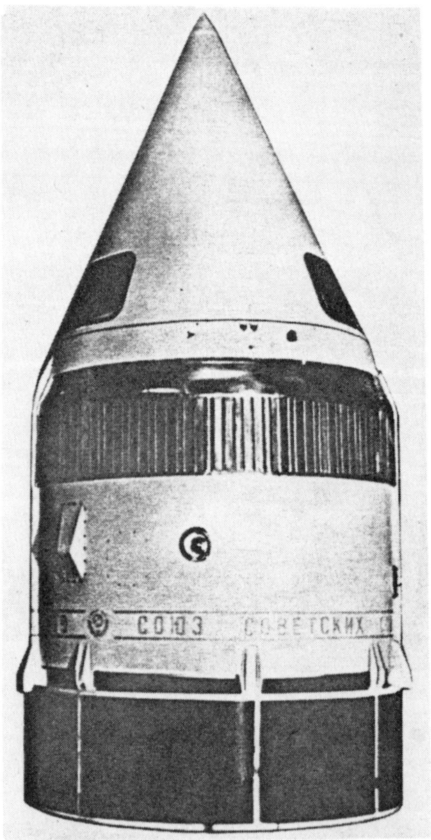

Figure 3 Luna 1 and last stage of launch vehicle. This final stage accompanied Luna 1 on the first flyby of the moon and released a sodium cloud visible from earth at a distance of 113,000 km.

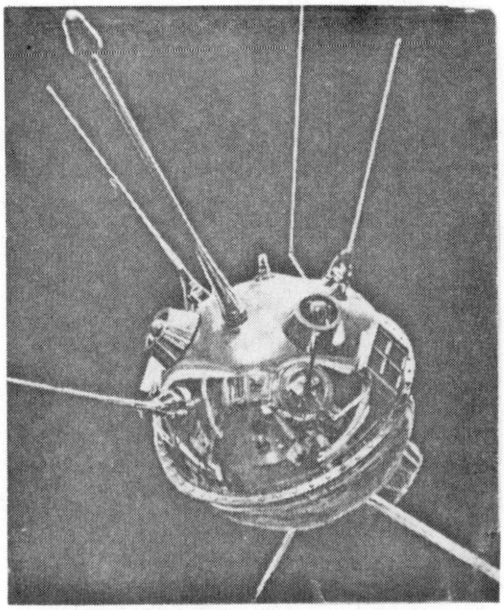

Figure 4

Luna 1 first man-made object to fly by the moon and enter a heliocentric orbit. Launched from earth on 2 January 1959

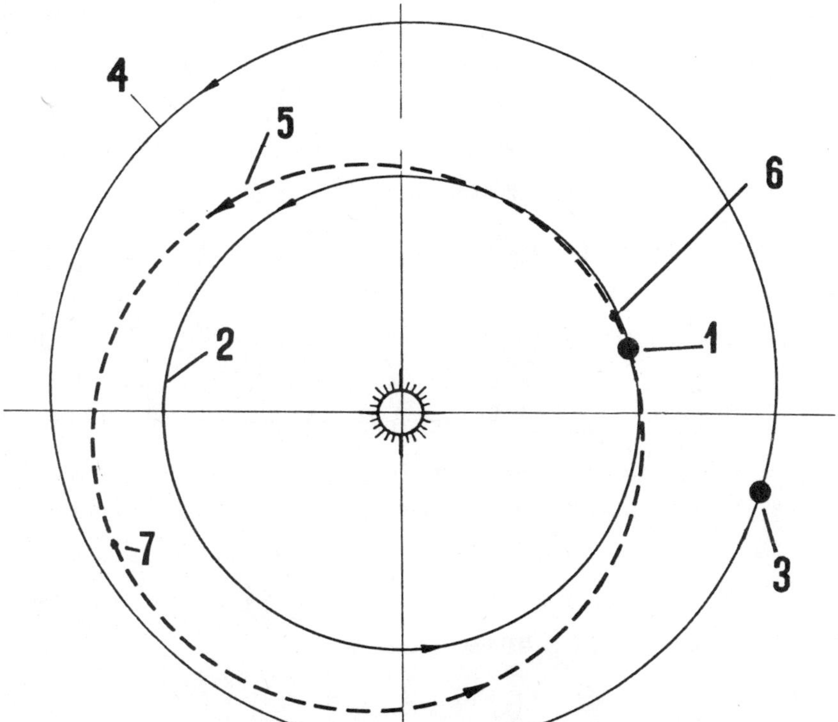

1. The earth
2. Orbit of the earth
3. Mars
4. Orbit of Mars
5. Trajectory of Luna 1
6. Perihelion of Luna 1 (146.4 million km)
7. Aphelion of Luna 1 (197.2 million km)

Figure 5 Trajectory of Luna 1 after its passage by the moon

4. Apparatus for the study of gas component of the interplanetary matter and corpuscular radiation of the sun
5. Instruments for measuring magnetic field of the earth and detection of the moon's magnetic field
6. Apparatus for the study of meteoric particles
7. Instrument for recording heavy nuclei in the primary cosmic radiation
8. Instruments for recording intensity and its variation of cosmic rays and photons in cosmic radiation.[3]

To measure magnetic fields a three-component magnetometer (SG-50) was employed. The device was located on the central rod extending slightly over one meter from one end of the probe. The results of this experiment showed a sharp decrease of the magnetic field as the probe receded from earth and an imperceptible lunar magnetic field.[4] A 0.2 m^2 ammonium phosphate piezoelectric detector recorded micrometeorites in three ranges: 2.5×10^{-9}g to 1.5×10^{-8}g, 1.5×10^{-8}g to 2×10^{-7}g, and greater than 2×10^{-7}g. Results of this experiment have been published.[5] Two gas-discharge counters, a scintillation counter, and ion traps were used to measure particle fluxes in the near-interplanetary space.[6]

The third and final stage of the A-1 launch vehicle (later christened the Vostok rocket) followed Luna 1 on its path toward the moon. This last stage was also equipped with scientific instruments. The major experiment on board called for the vaporization of 1 kg of sodium to form a visible cloud 100 km long at a distance of approximately 113,000 km from earth. The sodium was released on schedule at 0357 on 3 January 1959 Moscow time to the delight of Soviet scientists. The experiment was repeated and photographed nine months later during the Luna 2 mission. The Luna 1 final stage also carried instruments for measuring cosmic ray intensity and the rocket skin temperature.[7]

Soviet officials have maintained that Luna 1 was planned as a lunar flyby mission. However, many Western space specialists have concluded that Luna 1 was intended to impact the moon but missed. Sheldon even quotes Yuri Gagarin as admitting Luna 1 failed to hit the moon as planned.[8] The similarity of Luna 1 and Luna 2 and the fact that Luna 2 did strike the moon in September 1959 bring additional weight to this theory. From a scientific standpoint both types of missions, flyby and impact, have differing advantages. An impact mission can return data from closer to the lunar surface while a flyby probe can return data for a longer period of time and at post-lunar distances. From a technological standpoint, it is no harder to hit the moon than it is to pass the moon at a predetermined distance. In fact, during this same period the US attempted several lunar flyby missions with impact missions not beginning until 1962. However, without more convincing evidence to the contrary, the assumption will be made that Luna 1 was indeed intended to bypass the moon.[9]

LUNA 2

On 12 September 1959, three days before Premier Khrushchev was to start his whirlwind tour of the United States, the USSR launched Luna 2 on a direct trajectory to the moon. At *0002 hr 24 sec on 14 September Moscow time, Luna 2 became the first man-made object to land on another celestial body. It impacted on the lunar surface 30° N, 0° longitude, traveling with a velocity of 3.3 km/sec and at an angle of 60° to the horizontal. This momentous event was witnessed by observatories around the world when the spacecraft and/or the last stage of the launch vehicle kicked up clouds of lunar dust on impact.[10]

Luna 2 weighed 390.2 kg and was virtually identical to Luna 1. The only apparent difference in the 120-cm diameter sphere was a slightly modified antenna housing for the more sensitive magnetometer. The program of scientific investigation was also similar to that of Luna 1. No lunar magnetic field or radiation belts were measured within the accuracy of the instruments (60 gamma) down to an altitude of 55 km above the lunar surface, but measurements were made of the earth's own radiation belts. Two independent Cherenkov counters and six gas-discharge counters were carried, resulting in the collection of data on the particle fluxes of nuclei and on fluxes of high-and low-energy electrons and photons. Particle traps on Luna 2 also revealed a region of ionized particles 10,000 km from the moon. Again a 0.2 m^2 piezoelectric micrometeor detector was employed, but with improved counting ability.[11]

---

* *Time is given in this book using the US military standard. The first two digits are hours, the second two digits minutes on a 24-hour basis. Unless otherwise indicated, time is always Moscow time.*

As with the Luna 1 rocket, the last stage of the Luna 2 A-1 launcher accompanied the spacecraft to the moon and impacted shortly after Luna 2. An extremely high degree of accuracy was needed to hit the moon since there was no means of course correction after the shut-down of the third stage. A sodium cloud was released by the final stage of the launcher late in the evening on 12 September at a distance of 156,000 km as had been done during the Luna 1 flight. Photographs of the expanding cloud were taken by an earth observatory tracking Luna 2 (Figure 6). The cloud was tracked for several minutes during which time it reached a stellar magnitude of 4.5 and a maximum diameter of about 640 km. The final stage of the carrier rocket also carried instruments to measure the intensity of cosmic radiation.[12]

To allay fears of contamination the Soviet Union reported that both the Luna 2 payload and the last stage of the Vostok launcher were sterilized prior to launching. Serious questions have been raised as to the quality of sterilization of the Soviet Union (see also Venus section of this book) and as to the possibility of sterilizing the entire final stage of any rocket. However, subsequent findings concerning the nature of the lunar surface and environment indicate that terrestrial contamination of the moon by automated spacecraft is highly unlikely.[13]

To commemorate the first landing of man-made object on the moon, Luna 2 carried two spheres, one 9 cm in diameter and the other 15 cm in diameter. These spheres were fashioned out of pennants bearing the coat of arms of the USSR (Figure 7).[14]

LUNA 3

On the second anniversary of the launching of the world's first artificial satellite, Sputnik 1, the USSR began a mission which will surely rank as one of the most astounding technological achievements of any nation, considering the state of the art at that time. Although unmanned landings on the moon and planets and manned landings on the moon are given more prestige, the flight of Luna 3 in the light of the technology of that day was nothing short of miraculous. Luna 3, only the sixth spacecraft launching of the Soviet Union, was designed to fly behind the moon, photograph its far side, and then develop and transmit the pictures back to earth.

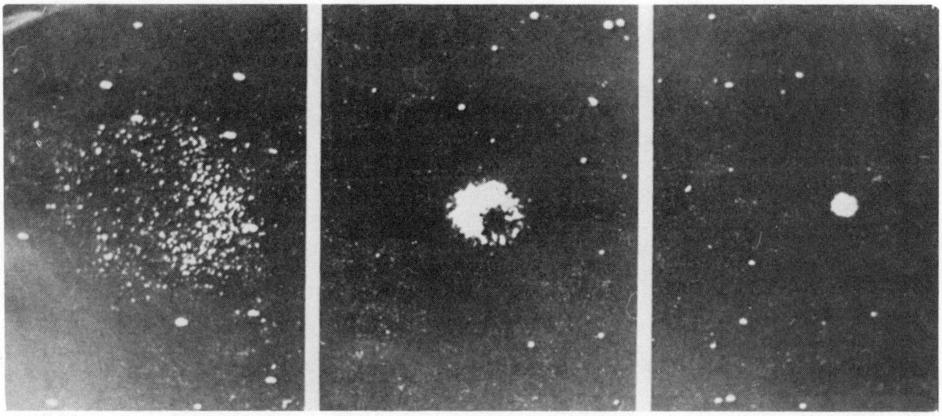

Figure 6a Three frames of film taken by an earth observatory of the sodium cloud released by Luna 2 on 12 September at a distance of 156,000 km

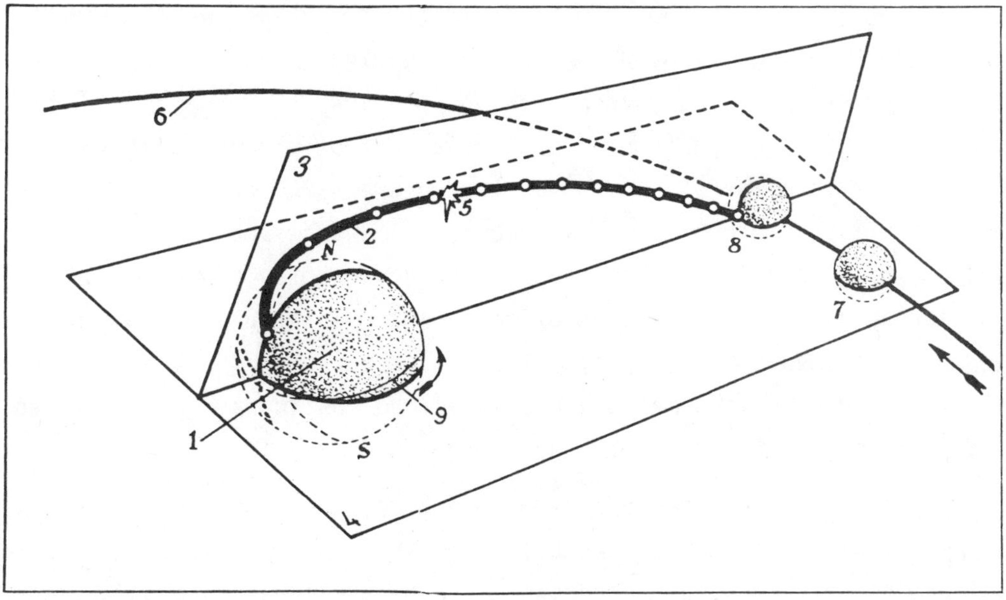

1. The earth
2. Trajectory of Luna 2
3. Plane of Luna 2 trajectory
4. Plane of lunar orbit
5. Location of sodium cloud formation
6. Orbit of moon
7. Position of moon at the launch of Luna 2
8. Impact of Luna 2 on the moon
9. Earth's equator

Figure 6b Trajectory of Luna 2 on its way to man's first direct encounter with the moon

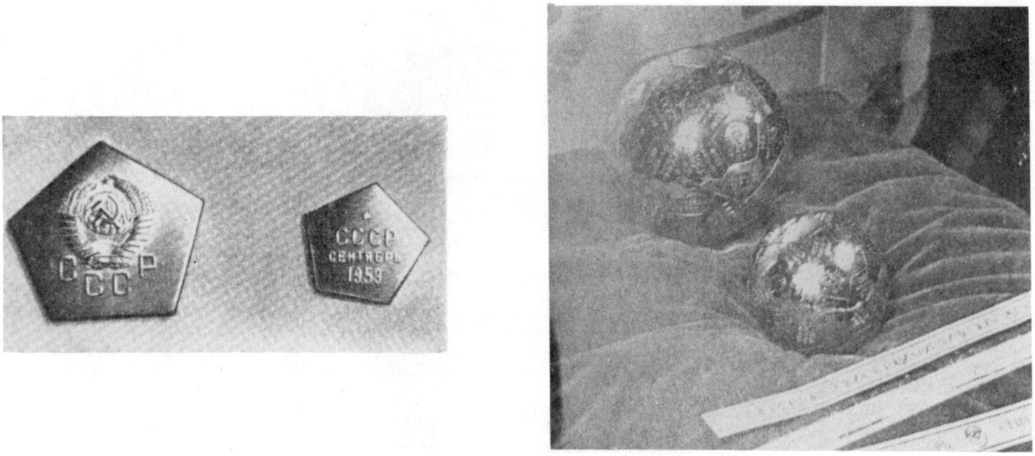

Figure 7 (Left) Sample of pennants used to form metallic spheres (right) taken to the moon by Luna 2 and the final stage of its launch vehicle to commemorate the landing of the first man-made object on the moon

The Luna 3 probe (also known as the Automatic Interplanetary Station, AIS) weighed 278.5 kg and measured 130 cm long and 120 cm across excluding antennae (Figure 8). In addition, the last stage of the carrier rocket contained 156.5 kg of equipment for a total weight of 435 kg of scientific payload. The interior of Luna 3 was kept between 25° and 30° C at a pressure of 1.5 atmospheres. The spacecraft was powered by chemical storage batteries which were subsequently recharged by blocks of solar cells surrounding the midsection of the craft. This enabled Luna 3 to be spin-stabilized during the flight from the earth to the moon. Instruments for recording micrometeors and solar and cosmic rays were carried as had been done with Luna 1 and 2.[15]

Although Luna 3 was not capable of making a course correction, it was equipped with an orientation system. A solar sensor was installed on the small end of the spacecraft opposite the camera windows through which a moon sensor could function. At 1716 hr on 6 October Moscow time, following a launch from earth two days earlier, Luna 3 made its closest approach to the lunar surface, passing at a distance of slightly more than 6000 km. On 7 October as it flew around the moon the craft's rotation was halted, and it was aligned along a sun-moon axis. Also in preparation for the photographic session the shutters covering the camera lenses were opened.

Figure 8a Replica of Luna 3 which returned man's first glimpse of the far side of the moon in October 1959

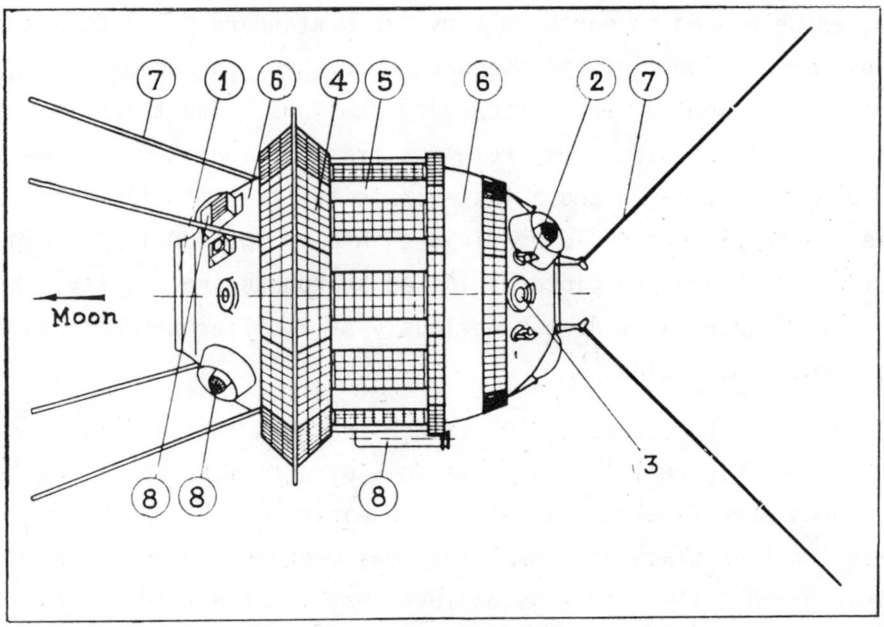

1. Viewport of camera system
2. Stabilization nozzle
3. Solar sensor
4. Solar cells
5. Shutters of the thermoregulation system
6. Thermal screens
7. Antennae
8. Scientific instruments

Figure 8b Diagram of Luna 3

At 0630 hr that day at a distance of 65,200 km above the surface, Luna 3 began taking pictures of approximately seventy percent of the heretofore unseen lunar far side. Forty minutes later, after taking at least 29 pictures, Luna 3 began rotating once more and continued its journey through space on a path which would once again take it near earth.[16]

The photographic system consisted of one camera with two lenses, one with a focal length of 200 mm to photograph the entire lunar disc and one with a focal length of 500 mm for more detailed photos. A special 35-mm isochrome film was used. This film was specially created to withstand the cosmic-ray bombardment it encountered between the earth and the moon. In earth-based tests the film had been exposed to 12 Mev electrons without adverse effects. After the pictures were taken, the film was developed and dried in a small automatic processor engineered to perform in a weightless environment. Development took about three minutes for each photograph.

The film also contained pre-exposed test patterns, some of which had already been developed on earth to serve as a standard for interpreting the lunar pictures. The finished pictures were then scanned by a TV camera and the signals sent to earth receiving stations. Two transmission rates were available: one slow-speed rate for transmission at distances of up to 470,000 km from earth and a faster rate for transmission near Luna 3's perigee. Each picture could be divided into up to 1000 lines (Figures 9 and 10). Although the pictures lacked sharpness and clarity, the topology of the lunar far side was obviously of a different character from the well-known near side.[17]

Luna 3 continued to apogee on 10 October and then to its first perigee on 18 October (Figure 11). There is some disagreement as to exactly when the pictures were finally transmitted to earth. Some authors imply that transmission took place at lunar distances whereas a planned retransmission near earth failed. Others believe that the transmission took place at approximately 40-50,000 km from earth at Luna 3's perigee. Soviet sources usually only admit to the possibility of transmission at both points and have, on occasion, listed either near the earth or near the moon as locations where actual transmissions took place.[18]

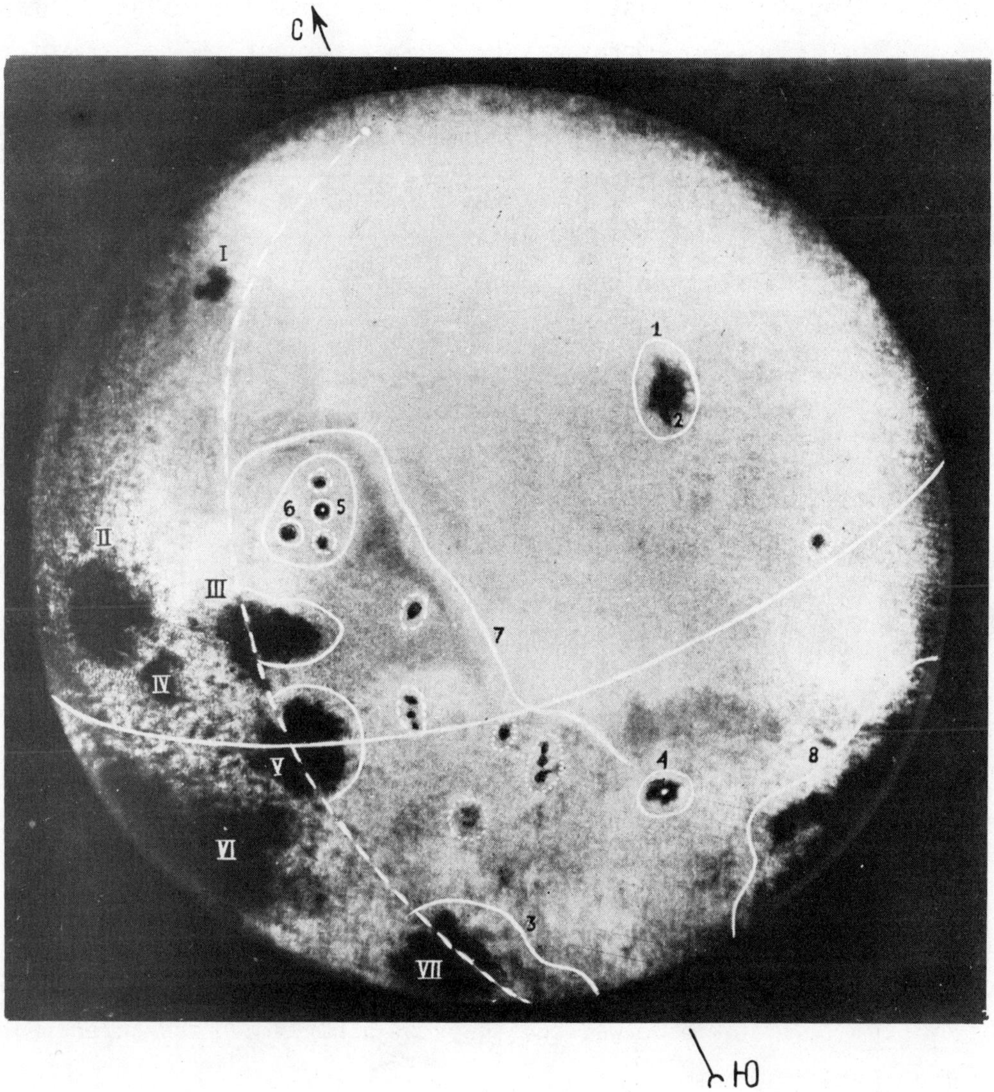

Far-side features:

1. Sea of Moscow
2. Gulf of Cosmonauts
3. Continuation of Mare Australe
4. Tsiolkovsky Crater
5. Lomonosov Crater
6. Joliot-Curie Crater
7. Sovietsky Mountain Range
8. Sea of Dreams

Near-side features:

I. Mare Humboldt
II. Mare Crisium
III. Mare Marginis
IV. Mare Undarum
V. Mare Smythii
VI. Mare Foecunditatis
VII. Mare Australe

Figure 9a Historic first photograph of the far side of the moon taken by Luna 3

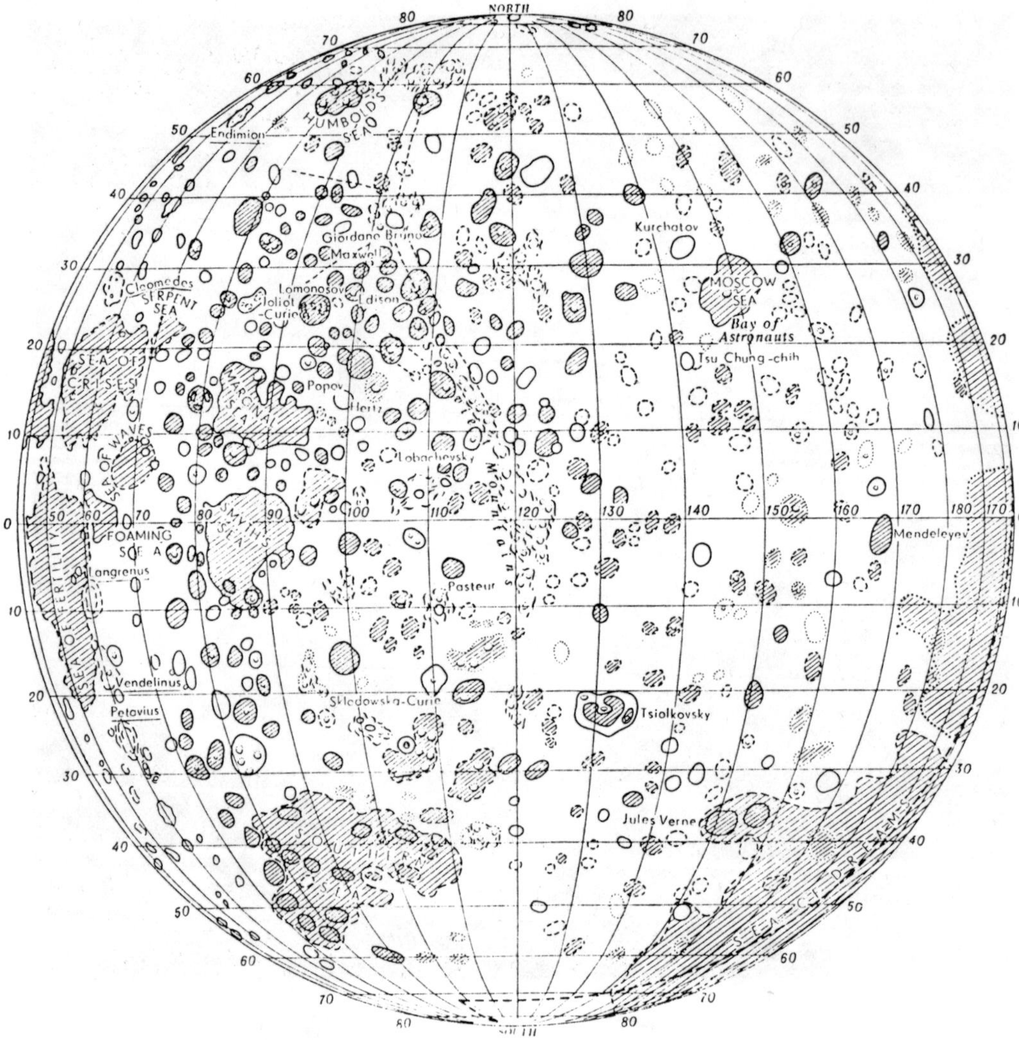

Figure 9b Map of lunar far side constructed with the aid of Luna 3 photographs

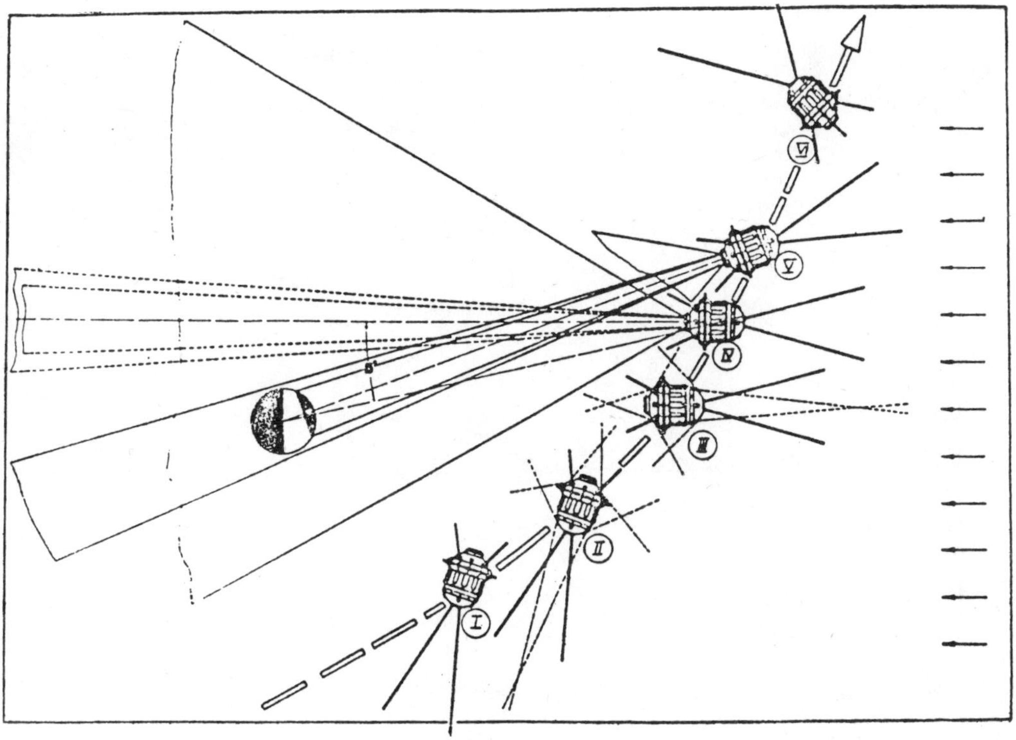

Figure 10   Orientation and sequence of Luna 3 photographic session

Communications with Luna 3 were abruptly lost in November 1959. From its calculated orbit, Luna 3 is generally thought to have been destroyed in the earth's atmosphere on 20 April 1960. However, Caidin has asserted that the automated spacecraft did not actually reenter until after 1962.[19]

Thus, in the span of only nine months the USSR had scored three impressive "firsts" in the fledgling lunar exploratory space program: first flyby, first impact, and the first photos of the far side of the moon. For the time being the Soviet Union was the uncontested leader of lunar exploration. However, for reasons only guessed at by Western analysts, the USSR did not launch another lunar probe until 1963 and was not successful in another lunar mission until January 1966.

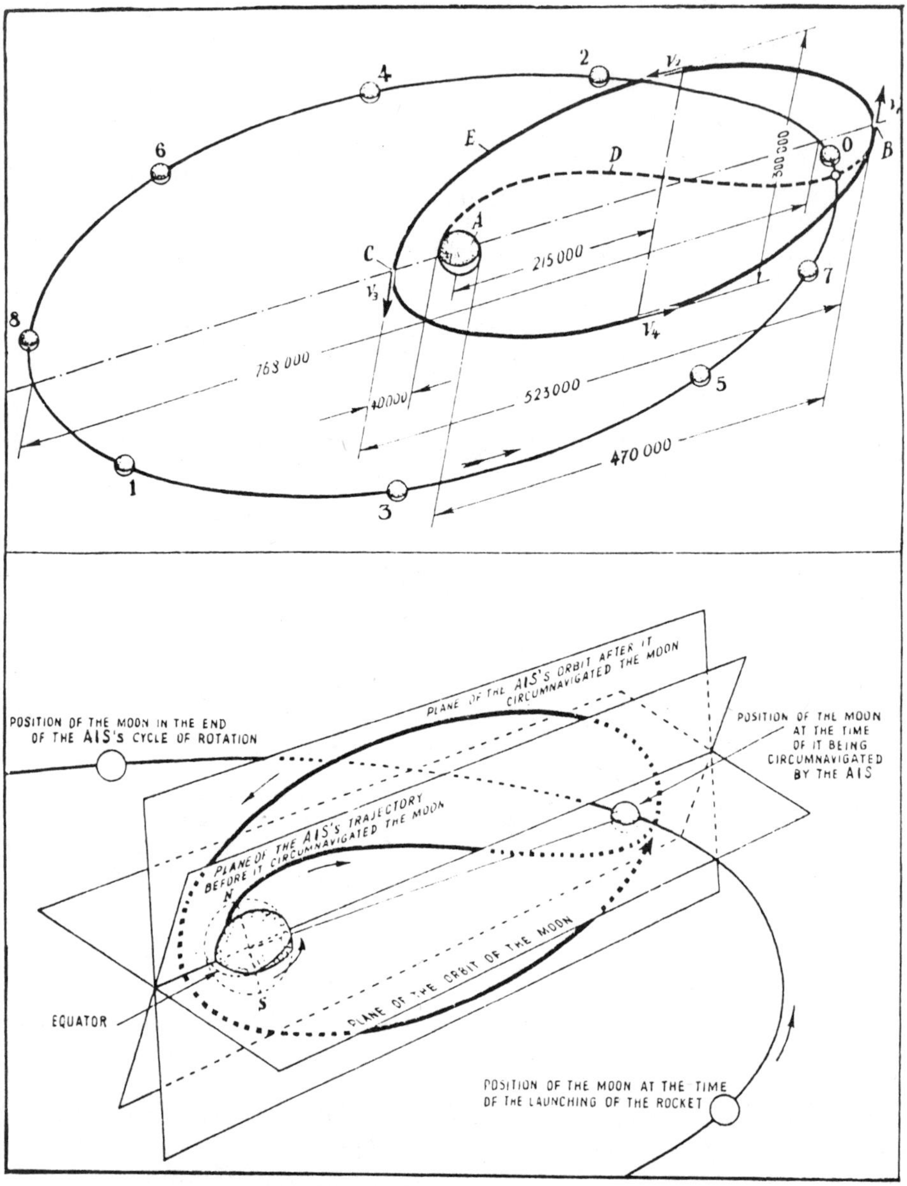

A. The earth
B. Apogee of Luna 3
C. Perigee of Luna 3
D. Initial Luna 3 trajectory
E. Final elliptical orbit of Luna 3
O. Position of the moon during Luna 3 encounter

V1. Minimum speed of Luna 3, 0.4 km/sec
V2. Luna 3 speed equal to 1.2 km/sec
V3. Maximum speed of Luna 3, 4.0 km/sec
V4. Luna 3 speed equal to 1.2 km/sec

Figure 11 Trajectory of Luna 3

Chapter 2

LUNA SERIES: SECOND-GENERATION SPACECRAFT

Lunas 1-3 had been launched atop the original version of the Vostok* rocket into a direct lunar trajectory. The maximum weight of these spacecraft was the combined weight of Luna 3 plus its final stage instruments bringing the total weight to 435 kg. To place heavier, more sophisticated payloads in the vicinity of the moon required the use of an upgraded space launch vehicle. A new version (A-2-e) of the original launch vehicle was first used during the 1960 and 1962 Mars attempts and during the 1961 and 1962 Venus missions. The up-rated launcher could place a much more massive lunar payload into earth orbit, and then refire, sending a robot probe into a translunar (or heliocentric) trajectory. Whereas the direct ascent first-generation Lunas did not have or need mid-course correction capabilities, most lunar probes fired from earth orbit have performed course corrections within a day of translunar injection.

The first attempt at using this launching technique appears to have been made on 4 January 1963. On that date an unmanned Soviet payload was placed into an earth orbit of 196 km by 167 km at an inclination of 65.0°. However, the lunar probe never left earth orbit and was destroyed reentering the atmosphere the following day.[20] Reports, unconfirmed by either Soviet or United States officials, of a second attempt during February have circulated. This probe reportedly never reached earth orbit, but fell into the Pacific Ocean shortly after takeoff.[21]

LUNA 4

Finally, on 2 April 1963 Luna 4 was placed into an earth orbit of 297 km by 167 km at 64.7°, a short while later fired out of earth orbit, and then placed on a path to the moon. Early in the morning of 6 April, Moscow time, Luna 4 skirted the surface of the moon at an altitude of 8500 km.

---

* *Vostok means "east" or "orient" in Russian*

Differing sources have reported that radio communications ceased some time between 6 and 15 April, depending on the source. The probe continued on its elongated earth orbit of 90,000 km by 700,000 km and was subsequently perturbed into a solar orbit.[22]

The Soviet Union has acknowledge that Luna 4 weighed 1422 kg, four times the weight of Luna 1. The exact mission of Luna 4, however, is not totally clear. Little is actually known of the fate of the mission, although some solar and cosmic-ray experiment results have been published. In many Soviet publications, Luna 4 has been linked with Lunas 5-9 which were admittedly soft-landing attempts. It is thus probable that Luna 4 was an effort to achieve the first lunar soft landing, but failed, in part, due to an unsuccessful mid-course maneuver. Hence, it is customary to believe Luna 4 was of the Luna 9 variety.[23] Had Luna 4 been successful, the Soviet Union would have moved a significant step closer to landing the first man on the moon.

One uncertainty of this assumption is the fact that Luna 5 did not fly for another two full years. An explanation of this apparent moratorium in that crucial stage of lunar exploration has been proposed by Phillip Clark. Since the introduction of the A-2-e launch vehicle in the Soviet Mars and Venus programs at the turn of the decade, the upper e-stage had performed dismally, stranding several expensive spacecraft in earth orbit. Clark hypothesizes that the Luna program (which is not restricted to infrequent launch windows like the Mars and Venus programs) was slowed down until the e-stage problems could be rectified.

KOSMOS 60

After the failure of Luna 4 in April 1963, no reports or firm rumors of any Soviet lunar attempts appeared for over a year. However, in June 1964 an article in an astronautics journal quoted sources as saying the USSR did attempt to launch a lunar probe in April 1964. Unfortunately the launch vehicle was reported to have failed within minutes of launch. No further independent confirmation of this article has been found.[24]

The next confirmed lunar attempt came nearly another year later on 12 March 1965. On that date, Kosmos 60 was placed in an earth orbit of 287 km by

201 km at an inclination of 64.7°. This time, however, the lunar probe failed to leave its orbital platform and decayed five days later.[25] It has been suggested that another attempt was made in April 1965, but did not reach earth orbit.[26]

LUNA 5

Finally on 9 May 1965 a satellite was placed in an earth orbit of 217 km by 151 km at 64.8°. With its subsequent injection into a lunar trajectory, the spacecraft was designated Luna 5. On 10 May, Luna 5 completed a mid-course maneuver which placed it on a collision course with the moon. This time there was no doubt--the first lunar soft landing was about to be attempted.[27]

The entire probe weighed 1476 kg with the landing capsule accounting for approximately 100 kg (see Luna 9 description). As the spacecraft approached the moon, an instrument malfunction prevented ignition of the retro-rocket at an altitude of 64 km. Luna 5 was destroyed upon hitting the moon at 31° S, 8° W at 2210 hr on 12 May. The Soviets had predicted touchdown at 2215 hr, a further indication of retro failure. A German observatory at Rodewisch reportedly was able to observe for about ten minutes a cloud of dust on the lunar surface 225 km long and 80.5 km wide at the site of impact.[28]

LUNA 6

When the next lunar launch window arrived a month after the launch of Luna 5, the Soviets were ready to try again. On 8 June 1965 a heavy satellite was placed on an earth orbit of 246 km by 167 km at 64.8° inclination. A short time later, the 1442-kg Luna 6 was fired into a lunar trajectory, during which all systems initially functioned properly. On 9 June, after a total of 12 communication sessions, Luna 6 was reoriented and its mid-course correction engine ignited to place it on an intercept with a predetermined region on the moon. However, the engine apparently failed to cut-off at the appropriate time. The additional impulse deflected the probe past the moon at a distance of approximately 161,000 km on 11 June and, as a consequence, Luna 6 continued into a heliocentric orbit.[29] Fate had once more denied the Soviets the goal they so desperately sought.

LUNA 7

Slightly more than a month after the launch of Luna 6, the USSR resumed its deep-space exploration with a combined lunar/planetary probe, *Zond 3. This spacecraft photographed the far side of the moon and then continued to the orbit of Mars. The apparent primary mission of the robot explorer was to test long-distance systems for future planetary spacecraft with the lunar far-side pictures offering an opportunity to evaluate the photographic and relay systems. The mission is described under the Zond and Mars programs of this book, but is mentioned here to provide the complete perspective of this era.

The Luna series of spacecraft resumed on 4 October 1965, the eighth anniversary of Sputnik 1. After leaving an earth orbit of 286 km by 129 km at 64.8°, Luna 7 began a three-and-one-half day trek to the moon.[30] After a successful course correction on 5 October, Luna 7 approached the moon in the early morning hours of 8 October. At approximately 0102 hr, Moscow time, the 1506-kg spacecraft ignited its rockets, but unluckily the probe was above the preprogrammed altitude. When the engine cut off, Luna 7 was still too high. Thus it struck the lunar surface with excess velocity at 0108 hr 24 sec near 9° N, 49° W.[31] Western observers reported from Moscow on 10 October that Luna 7 may have survived on the lunar surface for a short while, but these reports were never confirmed and Luna 7 is assumed to have perished on impact.[32]

LUNA 8

On 3 December the fifth attempt in 1965 to soft-land an automated capsule on the moon was begun. An orbiting platform carrying Luna 8 was placed that day in an earth orbit of 221 km by 181 km. However, this time its orbital inclination was 51.9° instead of the approximate 64.8° inclination of orbits of Lunas 4-7 and Kosmos 60. Apparently this was done to increase slightly the payload weight of the lunar probe, since by launching in a more easterly direction from Tyuratam the rocket gained more assistance from the rotation of the earth itself. All subsequent probes in the Luna series have used this inclination.

---

* *Zond means "probe" ("sond" or "sounding")*

Shortly after entering orbit, Luna 8, weighing a record 1552 kg, began its journey to earth's only natural satellite. The spacecraft performed flawlessly during its first 83-hour trip to the moon. Unfortunately, whereas Luna 7's retro-engine had fired at too high an altitude, Luna 8's engine now fired too late at a lower altitude than planned. As a result Luna 8 was still traveling at a high rate of speed when it struck the surface of the moon at 0051 hr 30 sec on 7 December west of Kepler crater at 9°8' N, 63°18' W. All signals ceased upon impact and the soft-landing capsule is presumed to have been destroyed.[33]

LUNA 9

Thus 1965 came to a close with a total of five unsuccessful Soviet attempts to place a surviving payload on the lunar surface. Meanwhile the United States was finishing the preparations on its first attempt to soft-land on the moon. If the Soviets were to win this portion of the space race, little time remained for additional attempts. Then at 1441 hr 37 sec on 31 January 1966, the Soviet Union launched the first lunar probe of the new year. The orbital launch platform with Luna 9 attached entered an earth orbit of 219 km by 167 km at 51.9°. Before completing even one orbit, the fourth stage of the launch vehicle ignited, propelling the Luna 9 spacecraft to a rendezvous with the moon some 3.5 days later (Figures 12 and 13). On 1 February at a distance of 233,000 km from the moon the probe's position was calculated, the craft was oriented, and a midcourse correction burn of 48 seconds was performed at 2229 hr. For the next two days Luna 9 hurtled moonward, rotating at 4 degrees per second to ensure proper temperature control and influenced only by the gravitational fields of the earth and later the moon.[34]

Lunas 4-8 and Kosmos 60 are believed to be virtually identical to Luna 9. The entire spacecraft after insertion into its translunar trajectory weighed 1583 kg of which the landing capsule accounted for about 100 kg. The remaining 1483 kg is reported to have been distributed as follows: lander equipment, 163 kg; instrument compartments jettisoned before landing, 300 kg; structure, 200 kg; and propellant, 800 kg. The overall length of Luna 9 was about 2.7 m. The actual landing capsule was egg-shaped with a diameter of about 58 cm. It was designed to be ejected from the descent

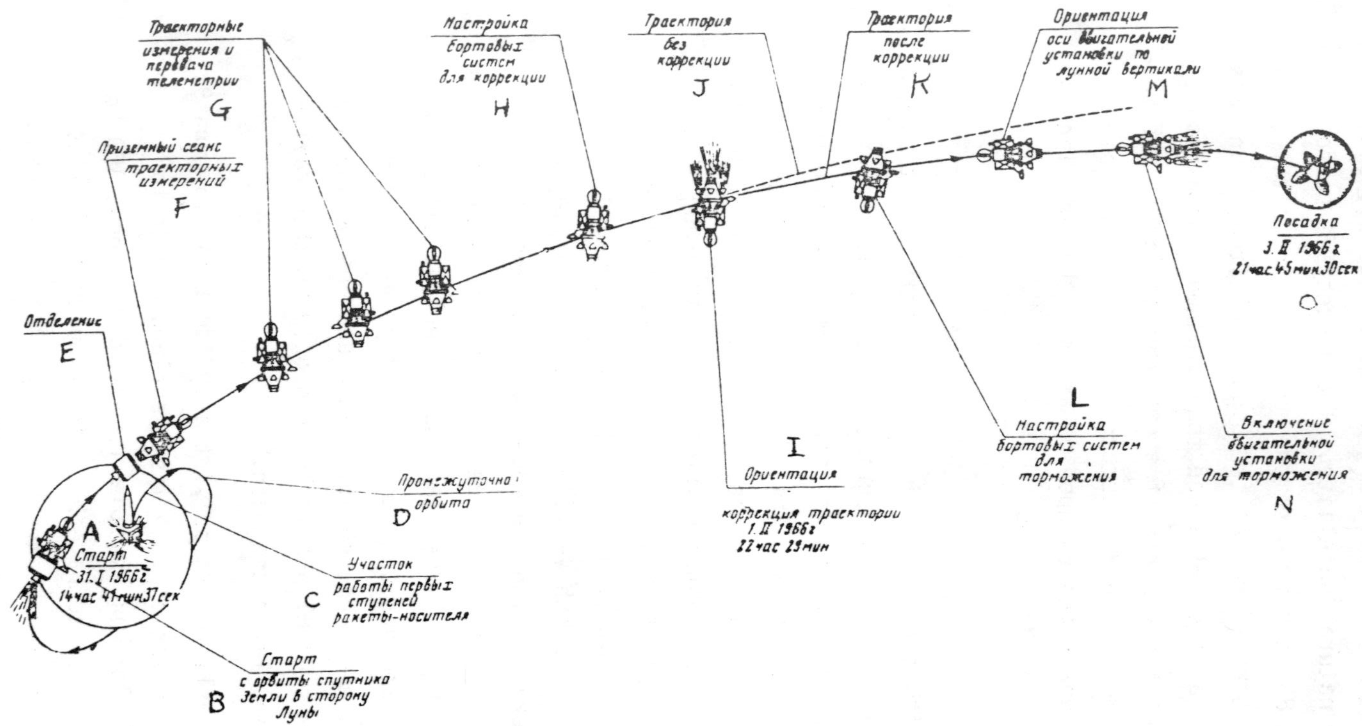

A. Launching, 31 Jan 1966, 14 hr 41 min 37 sec
B. Launching from earth satellite orbit in direction of moon
C. Segment of operation of first stages of carrier rocket
D. Intermediate orbit
E. Separation
F. Near-earth trajectory measurement contact
G. Trajectory measurements and telemetric transmission
H. Adjusting on-board systems for correction
I. Orientation. Trajectory correction 1 Feb 1966, 2229 hr
J. Trajectory without correction
K. Trajectory after correction
L. Adjustment of on-board systems for braking
M. Orientation of engine axis relative to lunar vertical
N. Starting engine for braking
O. Landing, 3 Feb 21 hr 45 min 30 sec

Fig. 12 Diagram of Flight of Lunar Rocket

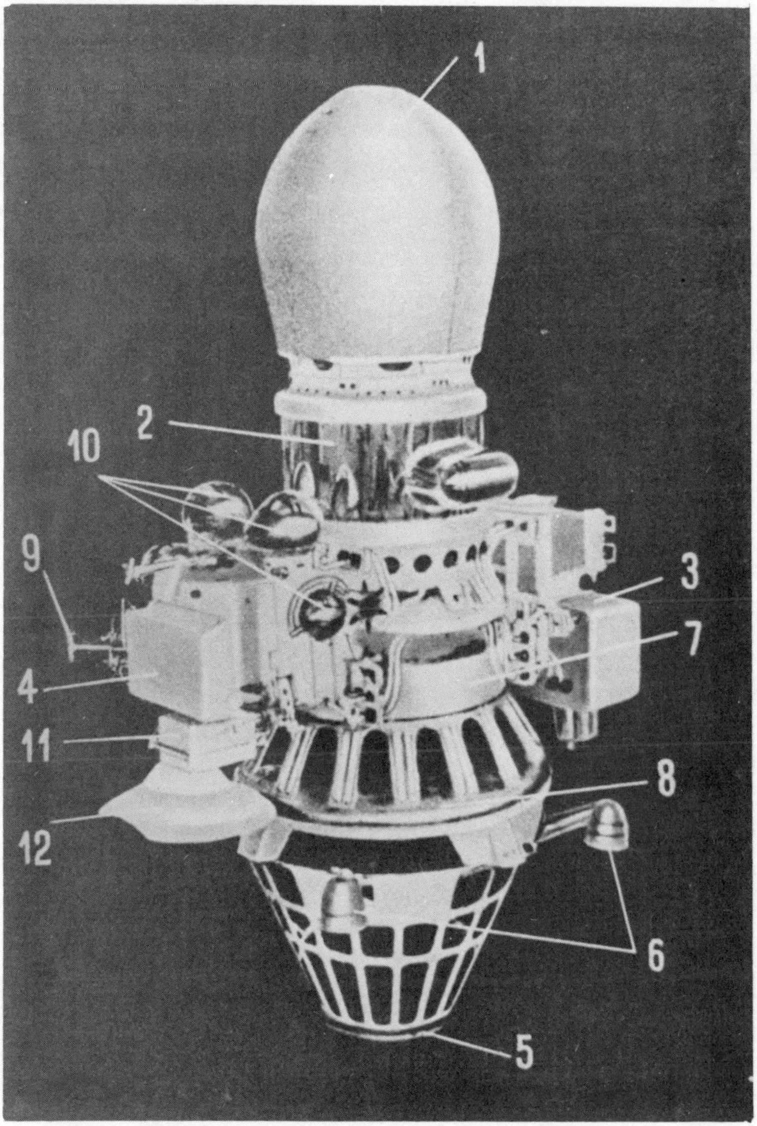

1. Landing capsule
2. Control section
3/4. Detachable sections with instrumentation
5. Liquid fuel engines
6. Control engines
7. Oxidizer tank
8. Propellant tank
9. Micromotor
10. Tanks with gas
11. Radioaltimeter
12. Antenna of radioaltimeter

Figure 13 Diagram of Luna 9

apparatus of the spacecraft immediately before the retro-engine made contact with the lunar surface. The landing capsule would then fall nearby on to the surface. The capsule's instruments were to be cushioned by interior shock absorbers. The four petal-like segments forming part of the capsule's enclosure, which also acted as part of the antenna system, would then spring open, uprighting and stabilizing the lunar station. Finally four 75 cm long antennae would be deployed for communications with earth receiving stations. The diameter across the opened petals on the lunar surface was about 160 cm. The height of the Luna 9 capsule including antennae reached about 112 cm (Figure 14).[35]

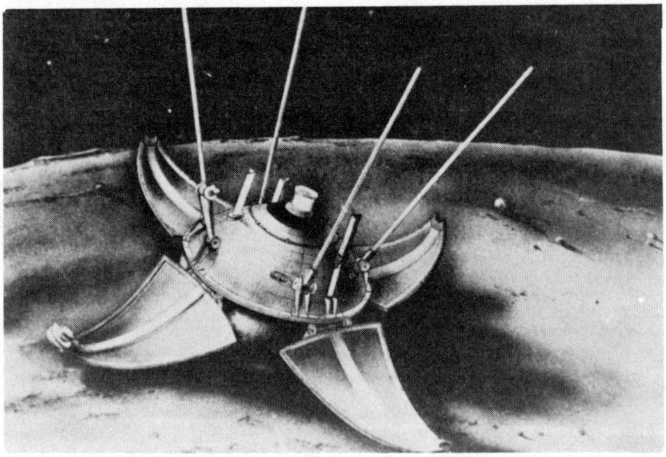

Figure 14 Sketch of Luna 9 on the lunar surface

The interior of the Luna 9 capsule was arranged on a shock-absorbing framework. The upper half of the capsule contained the television equipment and a 6-mm diameter, 10-mm long gas-discharge radiation counter. In the lower half were housed the chemical batteries, thermal control system, and the communications equipment. The thermal control system worked on the principle of evaporating water to cool the forced-circulated gas within the capsule whenever the interior temperature rose above a preset level.[36]

Communications were carried out via the petals of the capsule enclosure and the deployed whip antennae. During the flight and the first several minutes on the lunar surface the antennae concealed within the petals served to transmit messages to and receive messages from earth. After the petals opened, the four whip antennae served these functions. Suspended from all the whip antennae were brightness standards for calibrating pictures to the lunar surface and calculating the albedo of lunar rocks. Also mounted on the exterior of the upper half of the capsule were three dihedral mirrors to be used in obtaining steroscopic pictures of certain portions of the lunar surface.[37]

On either side of the mid-section of the complete Luna spacecraft were two detachable control compartments. One compartment housed the stellar orientation equipment whereas the other contained the radioaltimeter antenna and associated circuitry. These compartments were designed to be jettisoned from the craft moments before the retro-burn to minimize the weight to be decelerated. These compartment structures were so light that the atmospheric pressure within them had to be reduced as the Vostok launch vehicle climbed to lower pressures at higher altitudes.[38]

The above instrument compartments were attached to the framework surrounding the engine assembly. The engine assembly consisted of a spherical oxidizer tank, a toroid-shaped fuel tank (both made of alluminum alloy), and the combustion chamber. The engine assembly performed the two functions of course correction and spacecraft braking. Bottles of compressed nitrogen were used for the attitude-control system, while power for the spacecraft was supplied by chemical batteries, not the solar cells which are characteristic of Soviet planetary probes. Most of the batteries were housed in the detachable instrument compartments and were jettisoned immediately before the retro-burn.[39]

Approximately one hour before landing and 8300 km above the surface the Luna 9 spacecraft was oriented along the lunar vertical (Figure 15). All systems were thoroughly checked and the automatic landing sequence begun. When the radio-altimeter indicated an altitude of 75 km at 2144 hr 42 sec on 3 February, two commands were given. First, the detachable instrument compartments were jettisoned. Second, the retro-engine was ignited and the probe began decelerating from a velocity of 2.6 km/sec. A five-meter long probe ran from the mid-section of the spacecraft extending beyond the retro-engine for the purpose of marking the point of engine

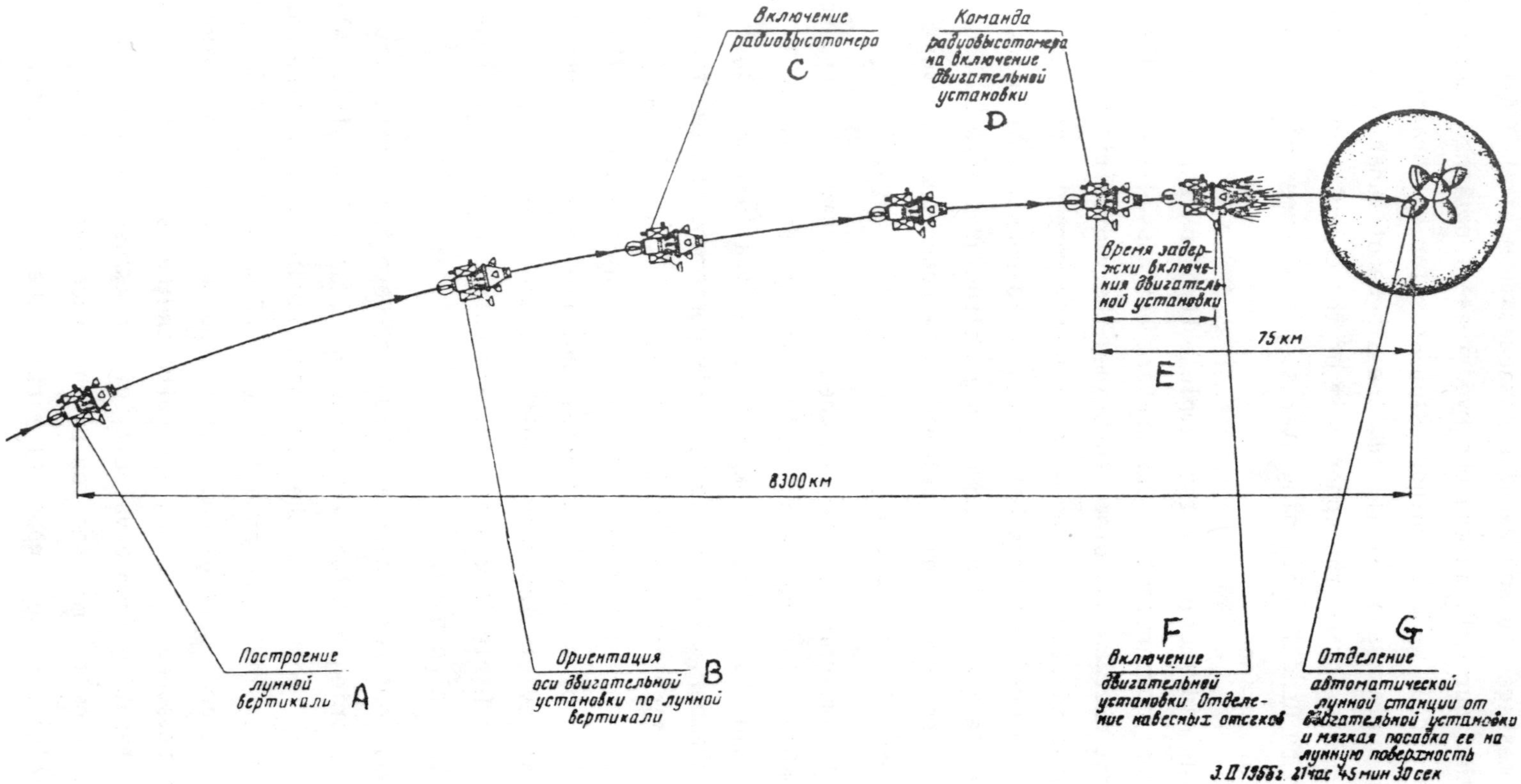

A. Construction of lunar vertical
B. Orientation of engine axis relative to lunar vertical
C. Activation of radioaltimeter
D. Radioaltimeter command for activating engine
E. Delay time in activating engine
F. Activating engine. Separation of attached compartments
G. Separation of automatic lunar station form engine and its soft landing on the lunar surface at 21 hr 45 min 30 sec on 3 Feb 1966

Figure 15a

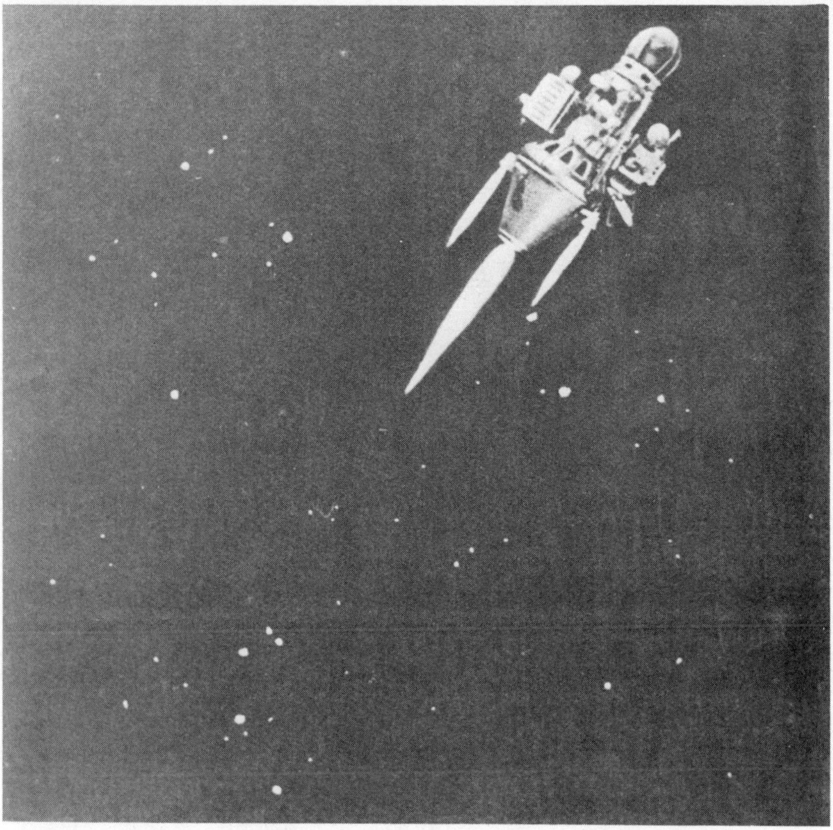

Figure 15b Luna 9 spacecraft braking before the release of the descent capsule

cutoff and ejection of the landing capsule. When the probe made contact with the lunar surface these operations were carried out, and the capsule fell to the ground near the impact of the engine assembly (Figure 16).[40]

Luna 9 finally came to rest at 2145 hr 30 sec at 7°08' N, 64°22' W, west of Reiner crater in Oceanus Procellarum. The less fortunate engine assembly crashed into the moon with a velocity of 5.5 - 6.0 m/sec.[41]

Approximately four minutes after landing on the moon, the Luna 9 capsule had opened and transmissions to earth had begun. The first panoramic view of the lunar surface was taken and transmitted about seven hours later. The television system consisted of a fixed lens and a mirror capable of rotating in two planes. The mirror would rotate in the vertical plane for each line and then shift slightly in the horizontal plane for the next line of the photo. Each panorama contained about 6000 lines

FLIGHT AND OPERATION OF THE LUNA-9 PROBE

No./Date/Time	Operations
1. *January 31* 15h.53m. 21 s.	Carrier-rocket's final stage shut off; lunar rocket is separated from final stage; time-programming device switched on.
2. 16h.46m.— 16h.56m.	First session in moonward phase of flight: trajectory measurements and reception of telemetric data. Switching on of receiving and transmitting radio system, ventilation system and telemetric units. Trajectory measurements (determination by radio-engineering means of distance, velocity and topocentric coordinates: elevation and azimuth angle.) Transmission of telemetric data containing scientific information and data on temperature and pressure in compartments of vehicle and on operation of instruments.
3. 17h.57m.	Vehicle put into spinning motion to ensure normal temperature conditions for all parts.
4. 18h.19m.— 18h.43m.	Trajectory measurements and reception of telemetric data; astro-navigation system switched on and performance checked.
5. 23h.15m.— 23h.25m.	Trajectory measurements and reception of telemetric information.
6. *February 1* 04h.41m.— 04.h.56m.	Trajectory measurements and reception of telemetric information.
7. 15h.36m.— 15h.52m.	Radio communication session. Same programme as during first session plus verification of on-board time.
8. 20h.09m.— 20h.42m.	Radio communication session. Adjustment of rocket's trajectory correction systems. Check on correct data input to time-programming device. Reception of telemetric information on condition of main systems.
9. 21h.16m.— 23h.14m.	Trajectory correction session. Verification of on-board time. Measurement of radial velocity. Rough and fine solar orientation. Opening of shutters of lunar sensor and orientation by Moon. Warming up of control system. Operation of engine in correction conditions. Automatic shutdown of engine by acceleration integrator. Measurement of radial velocity after shutdown of engine. Check on execution of operations and on condition of instruments during entire session. Rocket put into spin on its long axis.
10. *February 2* 01h.36m.— 01h.46m.	Trajectory measurements and reception of telemetric information.
11. 16h.16m.— 16h.31m.	Radio communication session, trajectory measurements, telemetric check-up and switching on the station's TV system for short-time test.
12. 22h.42m.— 22h.59m.	Trajectory measurements and reception of telemetric information.
13. *February 3* 01h.22m.— 01h.28m.	Trajectory measurements and reception of telemetric information.

Figure 16

FLIGHT AND OPERATION OF THE LUNA-9 PROBE (CONT'D)

No./Date/Time	*Operations*
14. 15h.34m.— 16h.31m.	Adjustment of deceleration systems. Check-up on condition of instruments and on correctness of data input into time-programming device.
15. 19h.28m.	Beginning of final pre-landing radio communication session devoted to deceleration. Instruments to control rocket during deceleration switched on. Verification of on-board time and of initial data relayed during previous session. Trajectory measurements. Turning of rocket. Orientation of one axis towards the Sun and of other towards the Moon. Orientation of optical sight towards the Earth. Construction of lunar vertical at some 8,300 kilometres from Moon's centre and switching on of gyroscopic control system.
	Continuous check of operation of all on-board telemetric systems. Signal from radio altimeter that pre-set altitude has been reached; retro-rockets switched on.
16. 21h.44m.42s.	Astro-orientation system switched off, and control transferred to autonomous gyroscopic system. Engine turned on, compartments jettisoned. Preparation of landing system for touchdown on lunar surface.
17. 21h.45m.30s.	Probe makes a soft landing on the Moon.
18. 21h.49m.40s.— 22h.06m.	First radio communication session with *Luna-9* after landing. Receiving and transmitting radio system and ventilation system switched on, as well as scientific and service instruments. Transmission of scientific data and telemetric check of on-board systems. Heat control system switched on. Petal and whip aerials unfolded. Wirelesses switched from petal to whip aerials.
19. *February 4* 00h.00m.— 00h.15m.	Second radio communication session.
20. 04h.50m.— 06h.37m.	Relaying of TV pictures of the first lunar panorama and telemetric information.
	TV systems switched on. Surveying of moonscape and transmission back to Earth of lunar surface panorama. Reception of telemetric data on condition and functioning of systems and instruments.
21. 17h.00m.— 19h.54m.	Transmission of second lunar panorama and telemetric information.
22. *February 5* 04h.00m.— 04h.04m.	Third radio communication session.
23. 19h.00m.— 20h.41m.	Relaying of third lunar panorama and telemetric information.
24. 21h.47m.— 21h.56m.	Fourth radio communication session.
25. 22h.28m.— 22h.45m.	Fifth radio communication session.
26. *February 6 and 7* 23h.37m.— 01h.55m.	Additional session. Transmission of pictures of individual areas of lunar surface and telemetric information.

Figure 16 (Cont'd)

with 500 elements per line. The picture covered an angular displacement of 18° down and 11° up, taking about 100 minutes to transmit (one line per second). The camera itself was already inclined at an angle of 16° to the horizontal. This ensured that the field of view would be directed at the lunar surface. The camera was a cylinder 8 cm in diameter and a little over 25 cm long. It weighed about 1.5 kg and only consumed 2.5 watts of power. Only a small portion of the camera containing the lens window and an upper solar screen protruded above the capsule. This set the lens window about 60 cm above the ground (Figure 17). That portion of the camera which did protrude was gold-plated to aid in the thermal control of the system.[42]

Focusing was designed for 1.5 m to infinity. At the 1.5 m distance (base of the probe) objects as small as 1.5 to 2mm could be distinguished (Figure 18). Between the transmission of the first and second panoramas and again between the second and the third, the Luna 9 capsule shifted on the lunar surface. This is thought to have been caused by settling in the lunar dust and/or the effect of the increasing temperature of the lunar day on the soil. This, fortunately, provided an opportunity to obtain better steroscopic pictures than originally designed. Luna 9 stopped transmitting at about 0155 hr Moscow time on 7 February when its batteries were depleted. A detailed analysis of the photographs taken has been published.[43] More important, however, the major question concerning the lunar surface had been answered: The soil was firm enough to support a 100-kg payload without noticeable effect. Future astronauts need have no fear of sinking into meters deep lunar dust.

KOSMOS 111

Less than a month after Luna 9 had made man's first soft landing on the moon, another lunar probe was launched from the Soviet Union. Kosmos 111 entered an earth orbit of 226 km by 191 km at 51.9° on 1 March 1966. However, the lunar spacecraft failed to leave earth orbit and was destroyed two days later upon reentering the atmosphere. Without further data it is impossible to determine definitely the mission of Kosmos 111. The probe may have been a lander of the Luna 9 type or may have been a lunar orbiter of the Luna 10 type, which was launched later that month. Most observers consider it to be of the latter variety.[44]

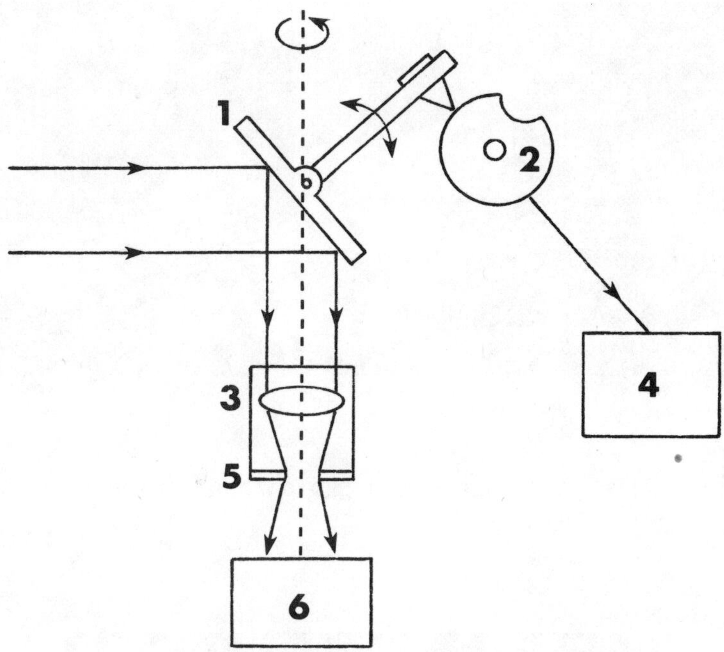

1. Mirror
2. Cam
3. Objective lens
4. Motor
5. Diaphragm
6. Light Receiver

A. Television Scanning System

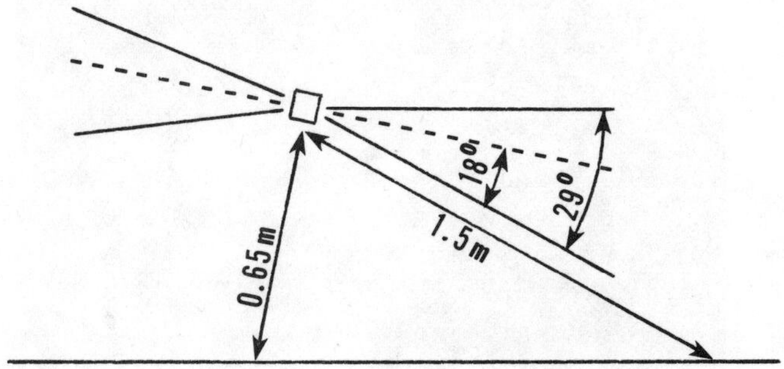

B. Orientation of Television Camera above the Lunar Surface

Figure 17

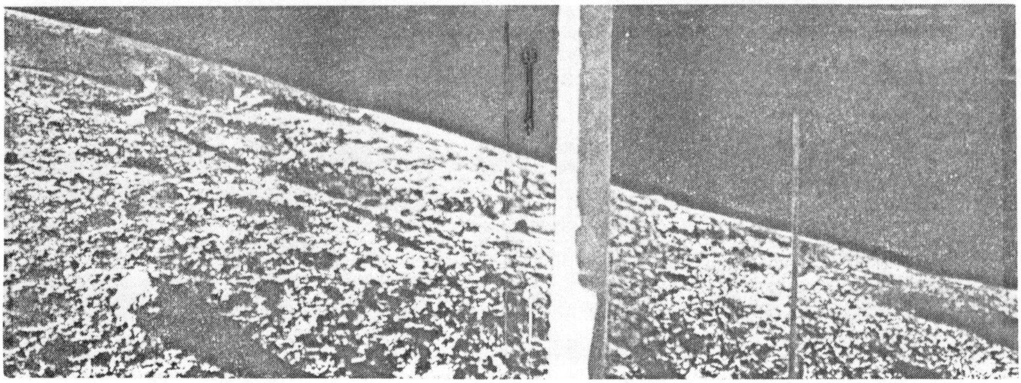

Figure 18a Portion of panoramic photograph of lunar surface taken by Luna 9

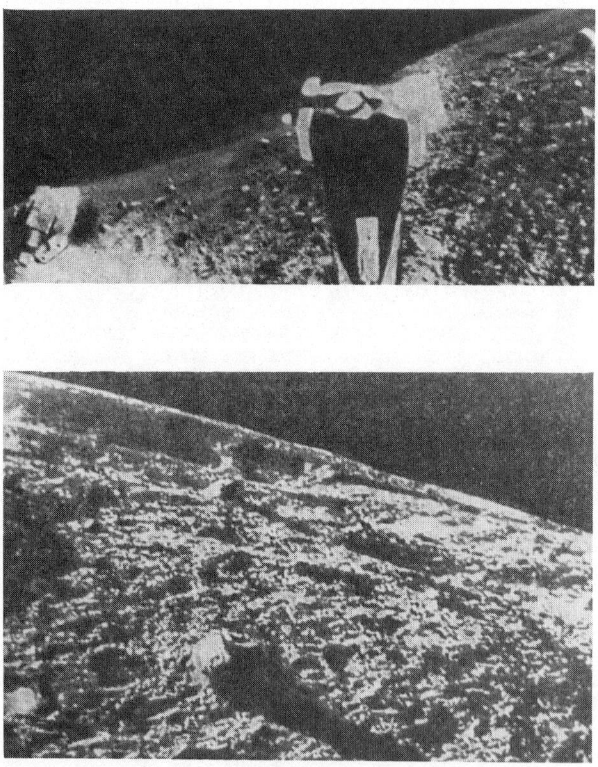

Figure 18b More photographs returned by Luna 9

LUNA 10

On 31 March 1966 Luna 10 was placed into a 250 km by 200 km earth orbit inclined 51.9° to the equator and subsequently fired into a lunar trajectory. As had been done in the past a mid-course correction was performed on 1 April, but this time the spacecraft was not directed to land on the moon. Luna 10 was on its way to becoming the first artificial satellite of the moon.[45]

After insertion into a translunar trajectory, the complete Luna 10 spacecraft weighed 1600 kg and was almost 4 m in length.[46] The engine assembly, instrument compartments, and control section were virtually the same as on Luna 9. This time, however, the landing capsule was replaced with a 245 kg lunar satellite (Figure 19). The roughly cylindrical

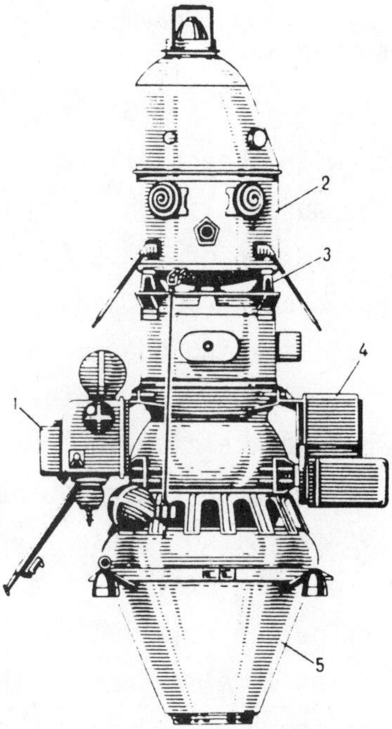

1. Radio equipment and scientific instruments
2. Luna 10 lunar satellite which separated from the remaining portion of the spacecraft after orbital insertion
3. Attachments between lunar satellite and the body of Luna 10
4. Astro-orientation system instruments
5. Spacecraft propulsion engine

Figure 19 Diagram of the Luna 10 spacecraft which placed the first lunar satellite in orbit around the moon

satellite was approximately 75 cm in diameter and 1.5 m long. The interior was maintained at a pressure of 850-860 mm Hg and at a temperature between 21° and 23° C. Again, as with Luna 9, power was supplied by batteries.[47]

Major experiments included:

1. Three component magnetometer for verifying lower limits of the possible magnetic field of the moon
2. Gamma-ray spectrometer intensity and spectral composition investigation of the lunar-surface gamma radiation
3. Counters for recording solar corpuscular and cosmic radiation, as well as for the investigation of low-energy electrons in order to determine the lunar ionosphere and study of charged particles in the "tail" of circumlunar magnetosphere of the earth
4. Ion traps for recording total ion and electron flux of solar wind and the search for lunar ionosphere
5. Piezoelectric pickups for recording in the interplanetary and lunar space of meteoric particles with mass over one hundred millionth of a gram
6. Infrared detector for determination of integral thermal radiation of the moon
7. Counters of low-energy X-ray photons for measuring X-ray fluorescent radiation of rocks on lunar surface.[48]

The three-component magnetometer was located at the end of a 1.5 m boom in order to isolate it from the fields associated with the satellite itself. This boom was folded along side the probe during launch. Magnetic fields up to 50 γ were detectable with a resolution of 1γ. The gamma-radiation detector compared the natural cosmic background radiation with that emitted by the lunar surface. By eliminating the amount of radiation induced by reactions with cosmic rays, the natural radioactivity, and hence the amount of uranium, thorium, and calcium-40, could be calculated. Five gas discharge counters were carried to record solar and cosmic protons and electrons as well as X-rays < 14 Å. Of these five, three were primarily designed to measure the fluorescence of lunar rocks due to solar X-radiation and, by doing so, determine the relative content of several elements in the rocks. Ion densities were measured with two four-electrode type ion traps and a modulation-type charged-particle trap. These would help determine the presence of a lunar ionosphere if it existed. A piezoelectric micrometeor detector similar to those on Lunas 1-3 had a registering area of 1.2 m^2 and a sensitivity of 7×10^{-8} g. Finally, an infrared detector consisting of two 15x30 mm flat plate receivers was used to measure heat fluxes emanating from the lunar surface. Apparently no photographic equipment was on board.[49]

In addition, two experiments could be analyzed on earth from the natural movement of Luna 10. By examining communications power levels as the lunar satellite passed behind the moon and reemerged on the other side, Soviet scientists could determine if any thin gaseous medium lay near the lunar surface and refracted the radio waves. Secondly, by measuring the natural changes in the orbit of Luna 10 the moon's gravitational field could be plotted. This was vitally important for the planning of future lunar missions due to the strong effect of mascons--mass concentrations in areas below the lunar surface.[50]

After a flight of three and one-half days and 8000 km from the moon the Luna 10 spacecraft was reoriented such that its rocket engine faced in the direction of travel. At 2144 hr on 3 April the retro-engine ignited reducing the velocity of the probe from 2.1 to 1.25 km/sec. (Figure 20 shows the Luna 10 trajectory.) Twenty

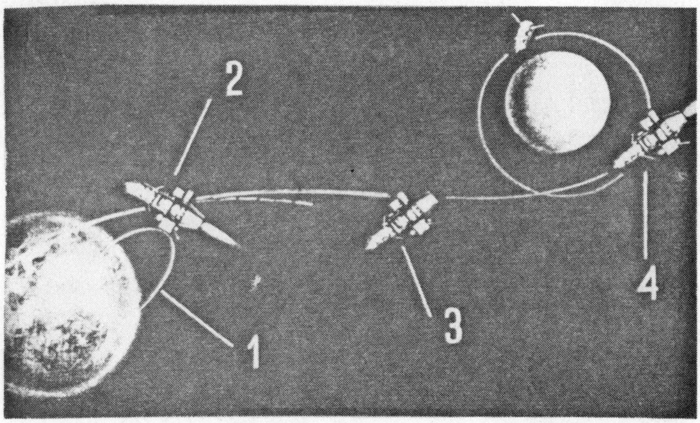

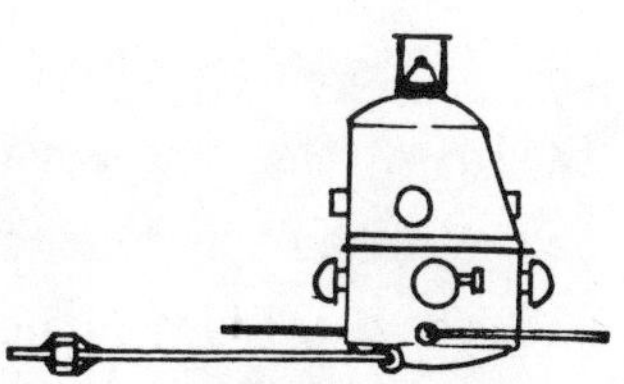

Figure 20 (Left) Trajectory of Luna 10 to the moon and entrance into lunar orbit

1. Initial earth orbit
2. Mid-course correction
3. Reorientation before braking
4. Lunar orbit insertion maneuver

(Right) Luna 10 lunar satellite as deployed in orbit about the moon

minutes after engine shutdown the Luna 10 satellite was ejected with a rotation rate of about 11° /sec. It entered a lunar orbit of 1017 km by 350 km at an inclination of 71° 54' to the lunar equator. In this orbit it completed one revolution every 2 hours 58 minutes 15 seconds. One of the first tasks performed by Luna 10 once in orbit was the playing of a portion of the Soviet anthem "Internationale" for the twenty-third Congress of the Communist Party.[51]

Luna 10 remained in orbit and continued to send back scientific information for 56 days. On 30 May 1966 after 460 revolutions, communications were lost when the chemical batteries finally were depleted. Its orbital parameters at that time were 985 km by 378 km at an inclination of 72°02'.[5]

LUNA 11

Three months after Luna 10 completed its program about the moon, another lunar satellite was launched from the USSR. Luna 11 was placed in a lunar trajectory on 24 August from an earth orbit of 201 km by 193 km at 51.8° after a 1103 hr launch time. After a course correction was made at 2202 hr on 26 August, Luna 11 then followed a flight profile similar to that used by Luna 10. The craft was oriented with respect to the lunar vertical at a distance of 8000 km from the moon. At 0049 hr on 28 August the retro-rockets were fired, and Luna 11 slipped into a lunar orbit of 163.5 km by 1193.6 km at an inclination of 27° with an orbital period of about three hours.[53]

Luna 11's weight has been quoted at 1640 kg, but pictures and other details of the spacecraft have been very scarce. Ambiguous data by the Soviets have led some Western analysts to place Luna 11 in the Luna 10 category whereas others think it was more similar to Luna 12.

When Luna 11 was launched, Soviet reports talked about the similarity of Luna 11 and Luna 10, although Luna 11 was described as slightly improved. The few construction details released fit Luna 10 and Luna 12 equally well. The initial description of the scientific experiments was virtually identical to that issued when Luna 10 was launched. In addition, Luna 11 performed no experiments of which Luna 10 was not capable.

Luna 11 entered a lunar orbit of inclination sufficiently different from that of Luna 10 to justify a repeat of the Luna 10 mission. The announced weights of Luna 10 and Luna 11 spacecraft were quite similar: 1600 kg for Luna 10 and 1649 kg for Luna 11. No mention was ever made of the weight of the actual Luna 11 satellite although the Luna 10 satellite was quoted at being 245 kg. Finally, no mention was ever made of any photographic capability of Luna 11 nor were any lunar pictures ever released.[54]

All the above arguments would lead one to believe that Luna 11 was a replica of Luna 10. However, other data suggest that Luna 11 was a forerunner of Luna 12. Most important is that Luna 11 was launched just two months before Luna 12, but five months after Luna 10. In fact, Luna 12 was launched only three weeks after Luna 11 had completed its program. This implies, considering past Soviet practices, that Luna 11 was scheduled to be the first Soviet lunar orbiter photographer, but that a malfunction prevented any picture returns. No weight has been released by Soviet sources of Luna 12, although a few Western observers have placed it at 1640 kg. This might imply that the Soviets did not want to disclose a weight equal to that of Luna 11, thus inviting comparisons and allegations of a failed Luna 11 photographer. Finally, Luna 14 is considered by most also to have been planned as a photographic lunar satellite. In this case, too, no mention has been made of a picture-taking mission, weight, or configuration.[55]

An indirect solution to this question may be found in a Soviet book titled, *Pokoryenye Kosmosa*. An illustration in this book suggests that Luna 11 was identical to Luna 10 except that the lunar satellite and propulsion unit remained attached in lunar orbit. The same figure also implies that the later Luna 12 and Luna 14 were sister spacecraft.

What is definite, though, is that Luna 11 continued to send back scientific data until 0503 hr on 1 October 1966 when the spacecraft's batteries gave out. During this time Luna 11 had completed no fewer than 277 revolutions about the moon.[56]

LUNA 12

On 22 October 1966 Luna 12 was rocketed into an earth orbit of 212 km by 199 km with an inclination of 51.9° and subsequently placed into a lunar trajectory. The objective of the Luna 12 mission was to obtain close-up photographs of the lunar surface, purportedly to examine possible future manned landing sites. The Soviets retained the basic engine and guidance systems used on Lunas 4-11 (Figure 21). However, the forward end of the

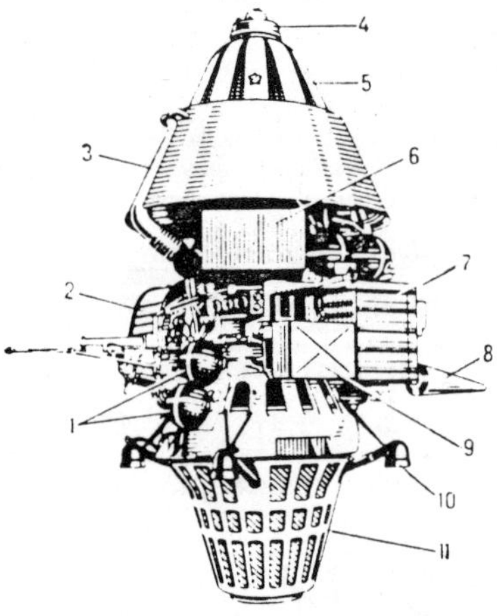

1—gas bottles of astro-orientation system actuating members; *2*—TV equipment; *3*—temperature control system radiator; *4*—radiometer; *5*—instruments compartment; *6*—chemical battery; *7*—optical and mechanical unit of astro-orientation system; *8*—antenna; *9*—electronic unit of astroorientation system; *10*—control rocket engines; *11*—retro-rocket engine for trajectory correction.

Figure 21 Diagram of Luna 12

probe was replaced with an enlarged instrument compartment, whereas the camera system was attached to the spacecraft body in the position of the detachable radioaltimeter of Luna 9. As mentioned previously, no weight was ever disclosed for the entire space vehicle, but the Soviets did announce that the weight of Luna 12 in lunar orbit was 1136 kg.[57]

On 25 October, after a mid-course maneuver two days earlier, as Luna 12 approached within 8000 km of the lunar surface, the craft was reoriented with respect to the lunar vertical as had been done with the previous eight Luna probes. At altitude of 1290 km and a velocity of 2.085 km/sec the spacecraft's liquid fuel retro-engine was fired for 28 seconds, reducing the probe's velocity to 1.148 km/sec and placing it in a 1740 km by 100 km lunar orbit with an inclination close to the equator. In this orbit Luna 12 made a complete revolution about the moon every three hours and twenty-five minutes.[58]

The photo-television system aboard Luna 12 was similar to that employed by Zond 3 in July 1975. Photographs were first taken by a conventional camera, then developed, and finally scanned by a television camera for transmission to earth. Each picture was composed of 1100 lines and covered a lunar surface area of approximately 25 km^2 from an altitude of 100-340 km. Resolution was reported to reveal craters as small as 15-20 m in diameter. Surprisingly very few pictures taken by Luna 12 have ever been released by the USSR. Apparently no photographs were taken by the orbiter until 29 October. This delay of four days in returning the first pictures is attributed by Shelton to an initial spacecraft orientation problem.[59]

In addition to the photographic mission, Luna 12 continued experiments begun by Luna 10 and Luna 11 in the areas of mapping the lunar gravitational field, establishing the micrometeor flux incident on the moon, measuring solar and cosmic radiation, and recording gamma and X-ray fluorescent radiation from the lunar surface. Finally, it has been reported that electric motors of the type to be used on the Lunokhod explorers were tested on board Luna 12. On 19 January 1967 after 602 revolutions Luna 12 ceased transmissions.[60]

LUNA 13

The year 1966 was one of the most successful years of lunar exploration for the USSR. January of that year saw the launching of the first spacecraft to soft-land on the moon. Exactly two months later, Luna 10 became the first lunar satellite. Two more satellites, one on a photographic mission, were placed in lunar orbit in August and October. As though the Soviet Union wanted to leave no doubt as to its determination and ingenuity in lunar exploration, a final lunar probe was launched in the waning days of the year. At 1317 hr on 21 December Luna 13 roared off a launching pad at Tyuratam, USSR, and into an earth orbit of 223 km by 171 km at 51.8°, before leaving on a 3.5-day journey to the moon.[61]

On 22 December a routine course correction was performed. At approximately 1900 hr on 24 December the craft was oriented, and at 2059 hr at a height of 70 km the retro-engine was ignited. Two minutes later, Luna 13 had successfully landed on the lunar surface 18°52' N, 60°0.3'W, some 400 km from the Luna 9 landing site. Four minutes after landing, the capsule opened and radio transmissions were begun.[62]

The outward appearance of the Luna-13 109-kg landing capsule was quite similar to that of Luna 9 (Figure 22). The main improvement was the

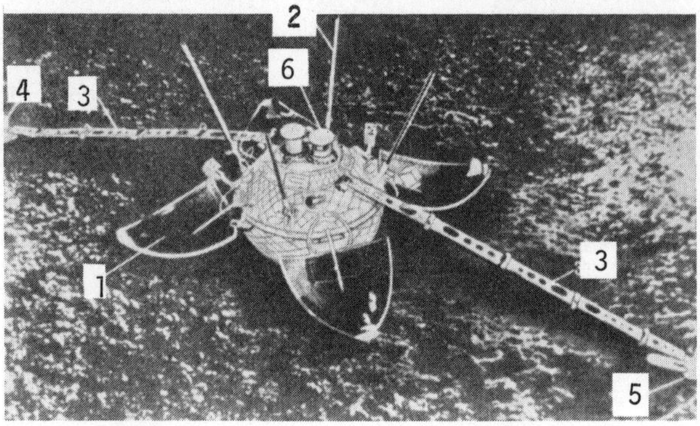

1. Petal-like cover with antenna
2. Rod antennae
3. Booms used to place scientific instruments on the lunar surface
4. Mechanical ground penetrater
5. Radiation density meter
6. Television camera

Figure 22 Sketch of Luna 13 on the lunar surface

addition of two folded booms, capable of extending 1.5 m beyond Luna 13. At the extremities of these arms were attached two scientific instruments. One contained a radiation densitometer for evaluating the composition of the lunar surface and the other arm contained a mechanical soil penetrater. These booms were deployed immediately after the opening of the capsule when the rod antennae had fully extended and the four lobe or petal-shaped antennae were two-thirds deployed.[63]

The mechanical strength of the soil was measured by a unique ground penetrometer (Figure 23). Within the device was an explosive charge which, when ignited, would force a 3.5-cm diameter titanium-pointed rod into the lunar surface. The explosive charge lasted for 0.6-1.0 sec and exerted a force of 5-7 kg. By measuring how fast and how far into the soil the probe penetrated, the nature of the soil could be determined. If the lunar soil was very dry and porous the indenter would reach its maximum extension of 5 cm very rapidly (0.003-0.05 sec). If the surface was more dense, the indenter would not penetrate so far and would take up to 0.5 sec. The explosive charge was fired at 2106 hr almost immediately after the penetrometer was placed on the lunar surface. Measurements showed that the indenter reached a depth of 4.5 cm, and as a result the lunar surface in the vicinity of Luna 13 was determined to be granular in nature

Figure 23a Photograph of lunar surface taken by Luna 13 with the ground penetrater and boom in the foreground

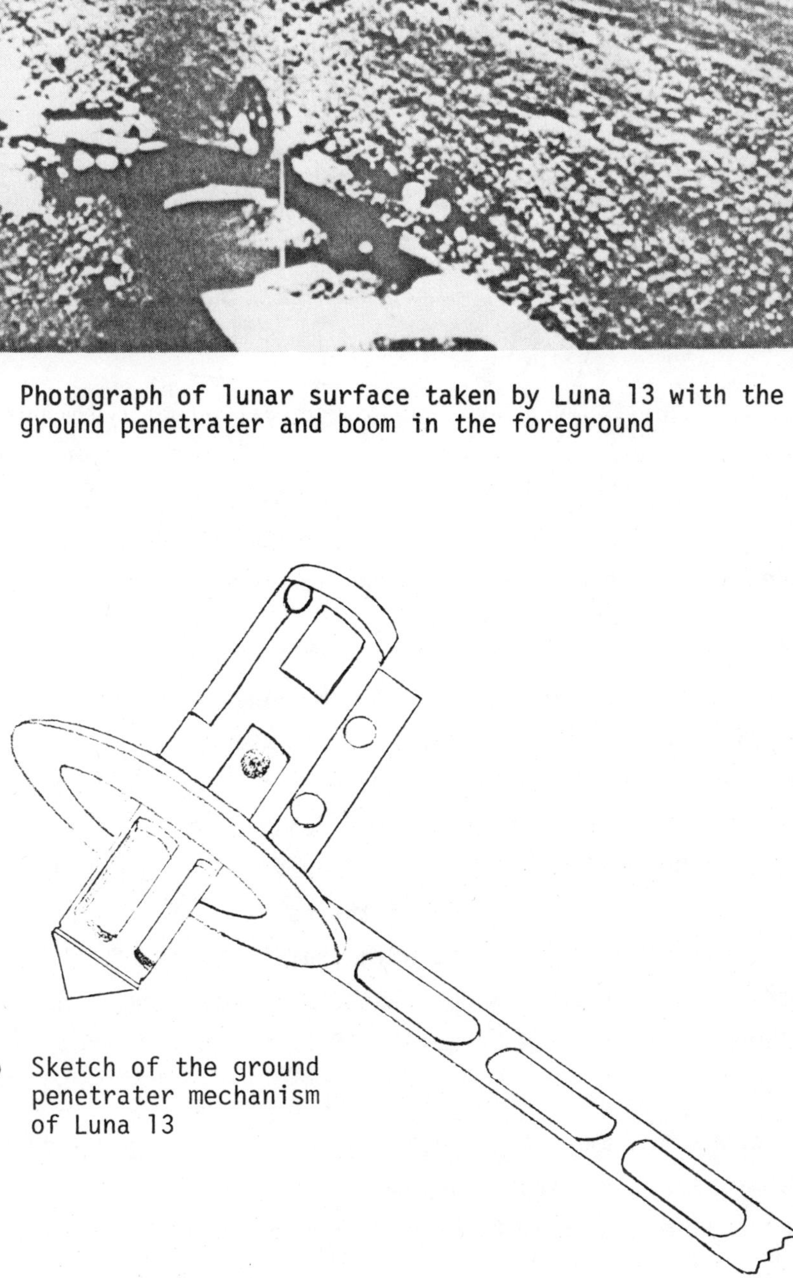

Figure 23b Sketch of the ground penetrater mechanism of Luna 13

and slightly cohesive with a density of about 0.8 g/cm^3. This corresponds in general with findings conducted on lunar samples returned from the later US Apollo and USSR Luna flights.[64]

The lunar soil density was also measured by a radiation densitometer located on the second extendable boom (Figure 24). The densitometer measured

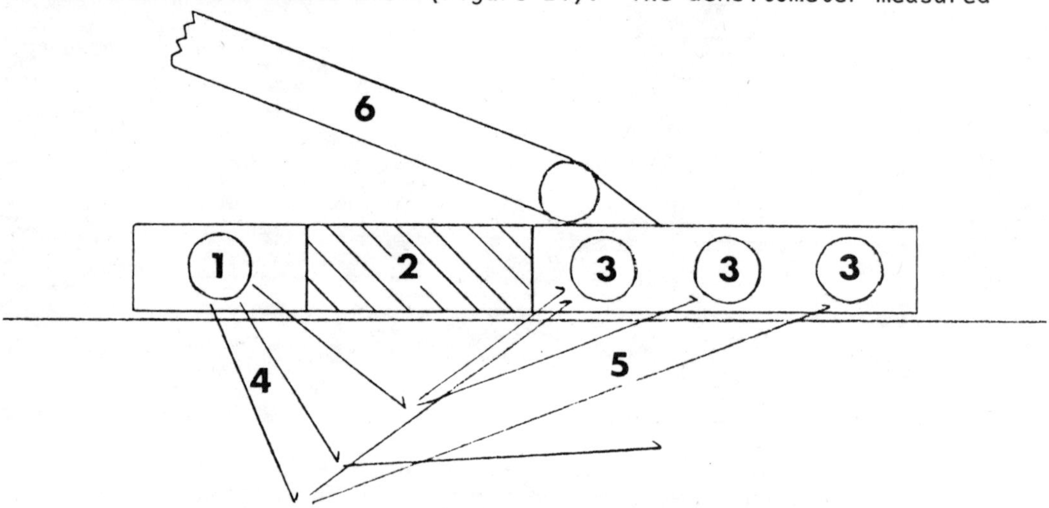

1. Gamma radiation source
2. Internal radiation shield
3. Gas discharge counters
4. Incident gamma radiation
5. Scattered radiation

Figure 24 Sketch of radiation density meter placed on the lunar surface by Luna 13

25.8 cm by 4.8 cm with a height of 1 cm, excluding two side-support plates and the attachments to the Luna 13 arm. Gamma quanta emitted by a cesium-137 sample were either absorbed or scattered by the lunar soil. By measuring the amount of scattering recorded by three independent pickups, an estimate of the density of the lunar soil could be made. A lead screen was located between the radioactive source and the pickups to shield the counters from direct radiation of the source. The results of this experiment corresponded very closely with that of the surface penetrater. To a depth of 15 cm the soil appeared to be loosely granular with a slight cohesiveness and density of about 0.8 g/cm^3.[65]

Other instruments were also carried on board the Luna 13 station. A piezoelectric dynamograph, consisting of three mutually perpendicular accelerometers, was used to determine the mechanical properties of the lunar surface down to a depth of 20-30 cm by measuring the duration and magnitude of the impulse resulting from the impact of Luna 13 with the surface during landing. Four radiometer pickups were employed to measure the thermal radiation of the lunar surface. By interpreting the data obtained with the sun at inclinations of up to 38° above the horizon, a "noon-day" effective radiation temperature of 117°C ± 3°C was calculated. In addition, a gas discharge instrument measuring direct and reflected cosmic particles determined that the lunar albedo (reflecting capacity) was about 25%.[66]

Of course Luna 13 also contained photographic equipment for visual exploration of the surrounding area. The camera system was virtually identical with that carried by Luna 9. The camera weighed 1.3 kg and consumed 2.5 watts. Tilted at an angle of 16° to the horizontal the television camera could resolve items as small as 1.5-2 mm at a distance of 1.5 m. As with Luna 9 a full panorama required approximately 100 minutes to scan (Figure 25). In one photograph the shadow of the Luna 13 station itself

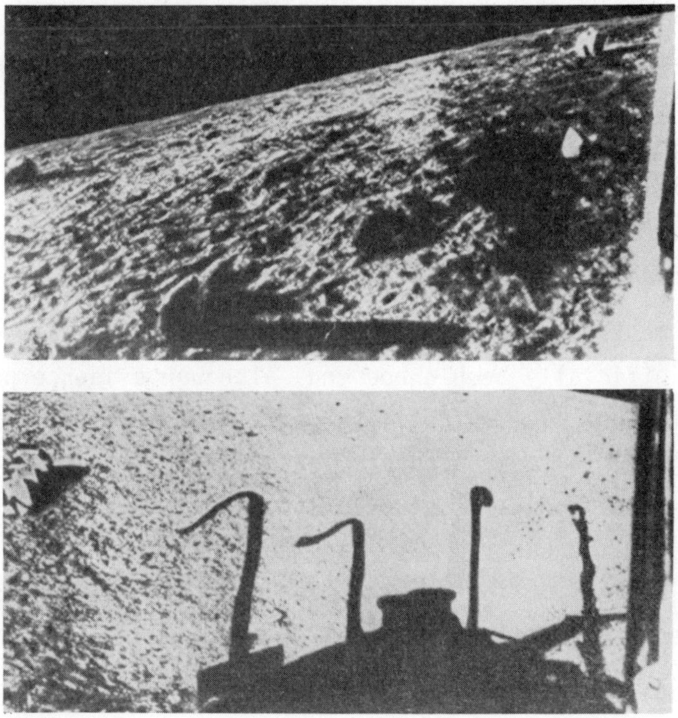

Figure 25 Photographs of the lunar surface taken by Luna 13. Note the shadow of the station itself in the lower picture.

is clearly visible, resembling some mythical, tenacled lunar creature. Examination of the photographs revealed no formations of diameter greater than 3.5 km within a radius of 100 km. Figure 26 presents a chart of major events and communication periods with Luna 13. On 30 December the scientific program of Luna 13 was completed as the batteries were exhausted.[67]

LUNA 14

After the stunning successes of 1966, it came as a mild surprise that the last lunar probe of the second generation of Lunas, Luna 14, was not launched until almost sixteen months after Luna 13. Although there was some reported activity surrounding Soviet moon shots in 1967 and early 1968 (see Zond series), no confirmed Luna series flights were attempted. Finally, on 7 April 1968, Luna 14 entered a temporary earth orbit of 242 km by 189 km with an inclination of 51.8°. Before completing one orbit, the last stage of the launch vehicle fired, hurling Luna 14 toward a rendezvous with the moon on 10 April. A mid-course maneuver was performed at 2227 hr on 8 April. Following the pattern of Lunas 10-12, Luna 14 reoriented itself in the vicinity of the moon and at a prescribed altitude fired its retro-engine. The spacecraft's velocity subsequently was reduced from 2.190 km/sec to 1.279 km/sec and the probe entered a lunar orbit of 870 km by 160 km at an inclination of 42° with an orbital period of two hours and 40 minutes.[68]

As previously mentioned (see Luna 11) the description and weight of Luna 14 have never officially been announced, although evidence points to its similarity with Luna 12. In fact, Tass gave only a general description of Luna 14's program: (1) more accurate measurement of the relative masses of the earth and the moon, (2) study of the stability of radio signals between the earth and Luna 14 with varying lunar positions, (3) investigation of the orbital motions of the moon, and (4) measurement of charged solar particles and cosmic rays.

Many Western analysts believe Luna 14 to be a failed TV orbiter of the Luna 12 variety, pointing to the lack of USSR enthusiasm for the flight and the lack of published scientific results. Luna 14 might even have been an afterthought for this second generation of Luna spacecraft, which

LUNA 13 TIME TABLE

Sequence No.	Date & Time	Operation
	24 December	
1	2101 hr	Automatic lunar probe Luna 13 made soft-landing
2	2105 hr 30sec	Reception of first signals from the lunar surface
3	2106 hr 2118 hr	First radio communication period Reception of telemetry data and instrument calibration
	25 December	
4	1515 hr	Period of transmission of television pictures and telementry data
	1653 hr	Switch-on of television equipment of the automatic lunar probe; scanning of lunar landscape and transmission of the first panoramic views of the lunar surface to earth; reception of telemetry data on the condition and operation of the on-board systems, scientific and servicing gear
	26 December	
5	1600 hr 1823 hr	Transmission of second lunar panorama and telemetry data
	27 December	
6	1648 hr 1955 hr	Transmission of third lunar panorama and telemetry data
7	2030 hr 2132 hr	Second radio communication period; telemetry data reception
	27-28 December	
8	2302 hr 0021 hr	Transmission of fourth lunar panorama and telemetry data
9	0041 hr 0106 hr	Third radio communication period
10	0223 hr 0248 hr	Fourth radio communication period
11	0705 hr 0913 hr	Transmission of fifth lunar panorama and telemetry data

Figure 26

substantiates Phillip Clark's suggestion that Luna 14 may have been Luna 12's backup spacecraft. In this regard, one source quotes Soviet officials as revealing that Luna 13 was to be the last of the Luna 9-13 type spacecraft. Reportedly, new unmanned Luna spacecraft would have mobile capabilities.[69] Again like Luna 12, Luna 14 is believed to have carried components of the electric motors to be used on the later *Lunokhods. No termination date of Luna 14 has been given, but the probe apparently lasted through April.[70]

---

* *Lunokhod in Russian means "moon walker" or "moon rover".*

Chapter 3

LUNA SERIES: THIRD-GENERATION SPACECRAFT

With the completion of the Luna 14 flight, the Soviet Union reached a plateau. The first-generation spacecraft, Luna 1-3, launched by the original Vostok launcher had explored near-moon space by flyby, impact, and photographic analysis. The second-generation spacecraft, Luna 4-14, launched by the augmented Vostok launcher had placed instruments on the surface of the moon and in lunar orbit. The next step for unmanned spacecraft would be the automatic return of a lunar soil sample and/or a vehicle on the moon capable of locomotion. Either spacecraft would represent a quantum jump in spacecraft sophistication and would necessarily weigh much more than previous probes. The Luna 1-3 spacecraft weighed between 361.3 kg and 435 kg whereas the Lunas 4-14 weighed between 1422 kg and 1640 kg. To send more massive machines to the moon would require a launch vehicle of greater strength. The Proton launcher had begun its career as a lunar launch vehicle in 1967 in accordance with the man-oriented Zond program. Now one year after Luna 14, it was to usher in a new dimension in space exploration.[71]

The first attempt to launch a Proton rocket mated with a probe of the Luna series is reported to have occurred in the first half of April 1969 (between April 5 and 15?). Possibly carrying a lunar sample returner, the rocket is believed to have exploded on a launch pad at Tyuratam. A second attempt on 14 June 1969 met with only slightly more success. This time the launch vehicle is believed to have left the launch pad but exploded in flight upon ignition of the second stage. If the Soviets had any intention of returning a lunar soil sample before the Apollo 11 astronauts returned from the moon, they only had one chance left.[72]

LUNA 15

On 13 July 1969, three days before the launch of Apollo 11, the first successful launch of the new generation of Luna probes carried Luna 15 into an earth orbit of 247 km by 182 km at an inclination 51.6°. A short time later the last stage of the carrier rocket propelled Luna 15 into a lunar trajectory. A mid-course correction was performed on 14 July amid world-wide speculation as to the probe's mission. It appeared as though the Soviets had pulled out all the stops and were determined to upstage or at least downgrade the imminent first manned lunar landing. At 1300 hr on 17 July, over four days after launch and with Apollo 11 fast on its heels, Luna 15 fired its retro-engine and slipped into a lunar orbit of 203 km by 55 km.[73]

Once in lunar orbit the mystery of Luna 15 only deepened. Two days passed before a brief announcement by Tass revealed that on 19 July Luna 15 had altered its orbit to 221 km by 9 km at 126°. If Luna 15 was intended to land on the lunar surface this appeared to be an unusual maneuver. On 20 July, only hours before the Apollo 11 Lunar Module was scheduled to land in Mare Tranquillitatis, Luna 15 made still another orbital correction to 110 km by 16 km at 127°. Now the spacecraft was in a more favorable position to attempt a landing. However, it was not until 1847 hr on 21 July, after the Americans, Armstrong and Aldrin, had become the first men to walk on the moon, that Luna 15 again fired its retro-engine and began the long descent to the lunar surface. In retrospect, with the exception of Luna 15's first orbital change, the spacecraft followed the three-four day lunar orbit pattern established by the later third-generation Luna landers. Four minutes after leaving orbit Luna 15 crashed into the Mare Crisium at 17°N, 60°E with a velocity of about 480 km/hr. After 52 revolutions about the moon Luna 15 had ended in the same mystery which had characterized its entire flight.[74]

Of course the USSR reported that Luna 15 had completed its designed mission. Although the probe represented a much more sophisticated spacecraft with a total launch weight of nearly 5800 kg by performing major orbital maneuvers, few observers believe Luna 15 was programmed to be merely a lunar satellite for four days and then destroy itself. Even now, eight years after the flight, there is little concrete evidence with which to classify Luna 15. Almost certainly Luna 15 was intended either to return a lunar sample before the Apollo 11 or to deliver the first Lunokhod rover. There are arguments to support both contentions.[75]

A Luna 16-type soil retriever would seem to be a logical attempt by the Soviets to "beat" the US to the moon. In addition, reports had been circulating in Moscow for several months of just such a flight. Finally, the next announced Luna mission was Luna 16 which indeed did return a small soil sample from Mare Fecunditatis. However, most physical evidence points to Luna 15 as being an unsuccessful forerunner of Luna 17 which deposited Lunokhod 1 on the lunar surface. Land and sea recovery teams needed to pick up a returning spacecraft from the moon were not deployed during the Luna 15 flight. The lunar orbits and landing conditions of Luna 15 were more similar to Luna 17 than Luna 16. There had also been a long history of Soviet intentions to develop a lunar rover. Whatever its mission, Luna 15 was soon forgotten in the euphoria which swept the world in the wake of the safe return of the Apollo 11 astronauts.[76]

KOSMOS 300 AND 305

Even though beaten by the United States in July 1969, the Soviets apparently were still very much intent on continuing their own expeditions. Two attempts in September and October 1969 were very probably made by additional Luna spacecraft of the Luna 16-17 variety. On 23 September Kosmos 300 was placed into an earth orbit of 208 km by 190 km with an inclination of 51.5°. The Luna probe apparently separated from the orbital platform, but either did not obtain sufficient velocity and merely moved into a different earth orbit, or escaped the earth's influence only to fail shortly afterwards. The probe may have been destroyed four days later upon reentering the atmosphere. Twenty-nine days later to the minute (almost exactly one lunar month) on 22 October Kosmos 305 was rocketed into space. Unfortunately the payload appears to have decayed before completing one complete revolution, although some debris remained aloft for two days. The year 1969 had begun with great expectations of a new generation of Luna spacecraft, but by the end of the year all five attempts had failed and of these only one, Luna 15, even reached the vicinity of the moon.[77]

A possible sixth attempt to launch one of the new heavier Lunas is occasionally referred to by Western observers. This launch failure was supposed to have occurred around 19 February 1970. Since it never reached orbit, very little is known of its purported mission.[78]

LUNA 16

The launch of Luna 16 on 12 September 1970 marked the beginning of a new series of outstanding achievements by Soviet space scientists. The last truly successful flight had been made in December 1966 by Luna 13. During this interval the nearly flawless progression of the US lunar exploration program overshadowed the technological advancements of the Soviets. The US had won the "space race" in July 1969 and despite occasional predictions of Soviet cosmonauts landing on the moon in the near future, the world public had for the most part begun to assume that due to internal social and economic problems the USSR had withdrawn from lunar exploration. Therefore, the automatic return of a lunar soil sample by Luna 16 caught the world by surprise, even though Soviet spokesmen had seriously spoken of lunar soil retrievers as early as 1966.[79]

Launched by a Proton rocket, Luna 16 entered an earth orbit of 241 km by 185 km at 51.5°, before speeding toward the moon 70 minutes later. One day after launch a routine course correction was accomplished with a 6.4-sec burn of Luna 16's main engine. On 17 September Luna 16 fired its retro-engine and entered an initial circular lunar orbit of 110 km with an inclination to the lunar equator of 70°. Two orbital maneuvers were performed on 18 and 19 September: The first moved Luna 16 into an elliptical orbit of 110 km by 15 km, the second altered the orbital inclination to 61° and slightly lowered the apolune to 106 km. As the probe approached perilune at 0812 hr on 20 September the braking engine was ignited and six minutes later Luna 16 had successfully soft-landed in darkness on the moon at 0°41'S, 56°18'E, after spending 75 hours and 26 minutes in lunar orbit. Luna 16 was now set to perform its primary task.[80]

In less than an hour after landing, at 0903 hr, an automatic drill began penetrating into the lunar surface for the purpose of extracting a soil sample. Seven minutes later at a depth of 35 cm the drill was stopped. The apparatus was then withdrawn, and the valuable cargo was carefully sealed in the return capsule. Its job completed, Luna 16 sat patiently on the moon awaiting the order to return to earth. At 1043 hr on 21 September the upper stage of Luna 16 blasted off the moon and began a three-day journey home. Three days later the Luna 16 capsule safely landed 80 km SE of Dzhezkazgan in the USSR at 0825 hr. By strictly automatic means the Soviet Union had returned 101 grams of lunar soil.[81]

A more detailed look at this remarkable spacecraft is warranted. Luna 16 weighed about 5800 kg at launch and stood approximately 4 m high with a base measurement of 4 m across from pad to pad (Figure 27). The craft is divided into two main sections: the descent stage and the ascent stage. The descent stage served the purposes of trajectory corrections along the path to the moon, orbital insertion, orbital maneuvers, descent and soft-landing on the lunar surface, collection of the lunar sample, and launch platform for the ascent stage. The ascent stage was designed to lift off the lunar surface carrying the lunar soil sample, perform any necessary course corrections (though none were ever needed), reenter the earth's atmosphere, and land intact so that the precious cargo might be recovered.[82]

The descent stage is constructed in modules according to the functions it must perform. The four large cylindrical tanks attached in pairs on either side of the descent stage contain the fuel needed for the trans-lunar segment of the flight and for the orbital insertion and correction burns. Shortly before the actual descent burn they are jettisoned along with components of the spacecraft orientation system no longer needed. A radioaltimeter provides rates of descent in the horizontal and vertical planes as well as height. Instruments measure temperature and radiation near the craft during the flight from earth and while in lunar orbit. They continue to function on the lunar surface after the ascent stage has left for as long as power remains. Power for all on-board systems is supplied by chemical storage batteries, giving the descent stage a life of only a few days on the surface. A temperature control system and an orientation sensing and control system were also employed, along with the necessary radio transmitters and receivers. Four compressible landing struts were used to cushion the impact of landing.[83]

When the spacecraft had been oriented in lunar orbit and was ready to begin the descent phase, the main engine was fired for a preset time. The probe then followed a free-fall path to the lunar surface (Figure 28).

Figure 27a Replica of Luna 16, the first fully automated spacecraft to land on the moon and to return to earth with a lunar soil sample (TASS from SOVFOTO)

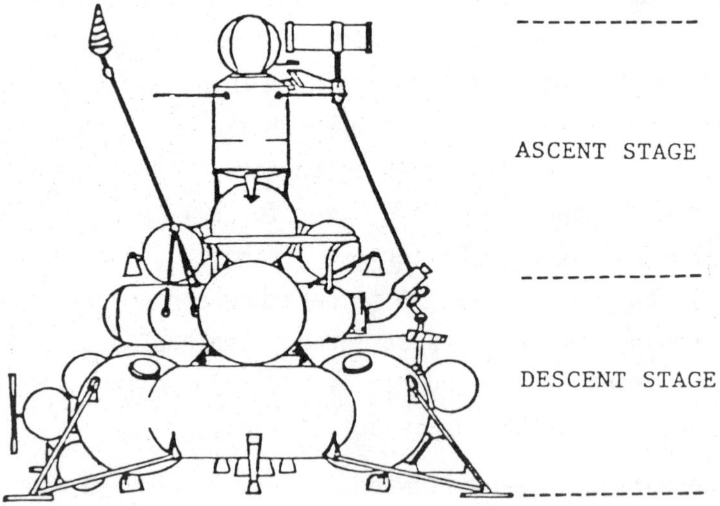

Figure 27b Luna 16 on the lunar surface after discarding the four large propellant tanks of the descent stage

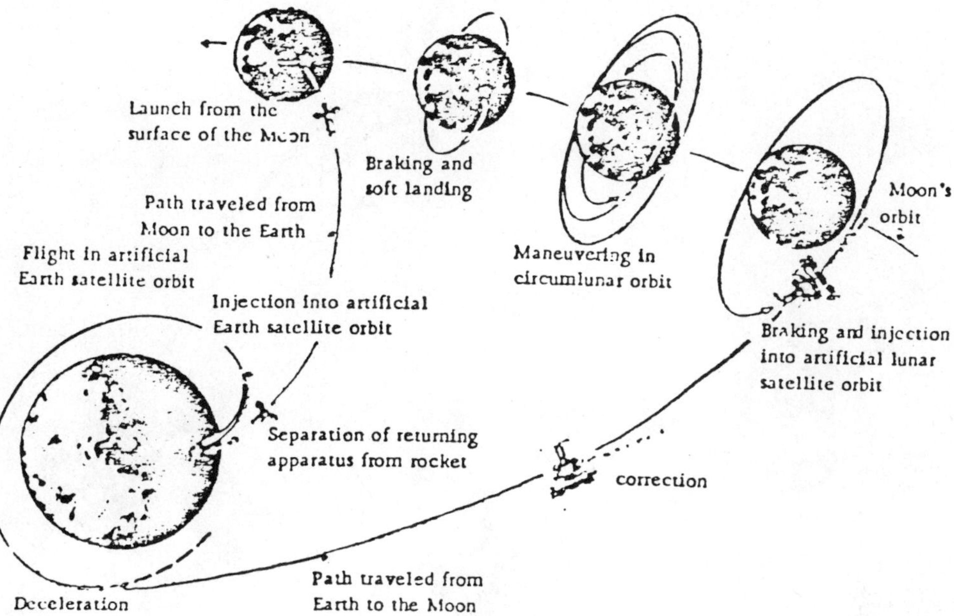

Figure 28a

At an altitude of 600 m the main engine and four vernier engines were again started and continued to burn at varying thrust levels as programmed by the on-board computers down to an altitude of 20 m and velocity of 2.5 m/sec. At this point two smaller engines took over until immediately before landing when they, too, shutdown. By switching to the two low-thrust engines, contamination of the lunar surface by the fuel exhaust could be greatly reduced. The weight of the entire Luna 16 spacecraft upon landing was 1880 kg.[84]

Once on the surface, two telephotometers arranged in horseshoe fashion on the descent stage were activated. Since the 0.9 m drilling arm had horizontal as well as vertical movement capabilities, study of the landing area by earthbound scientists permitted the drilling head to be programmed to avoid rocks or to investigate an unusual site. The drill itself was unlatched from its preflight position and rotated 180 degrees where a protective covering was ejected, exposing the drilling tube. After being positioned above the selected site, the drill head was lowered on to the surface, and the hollow rotating drill began penetrating the soil. During this operation, sensors within the drill measured the soil's resistance

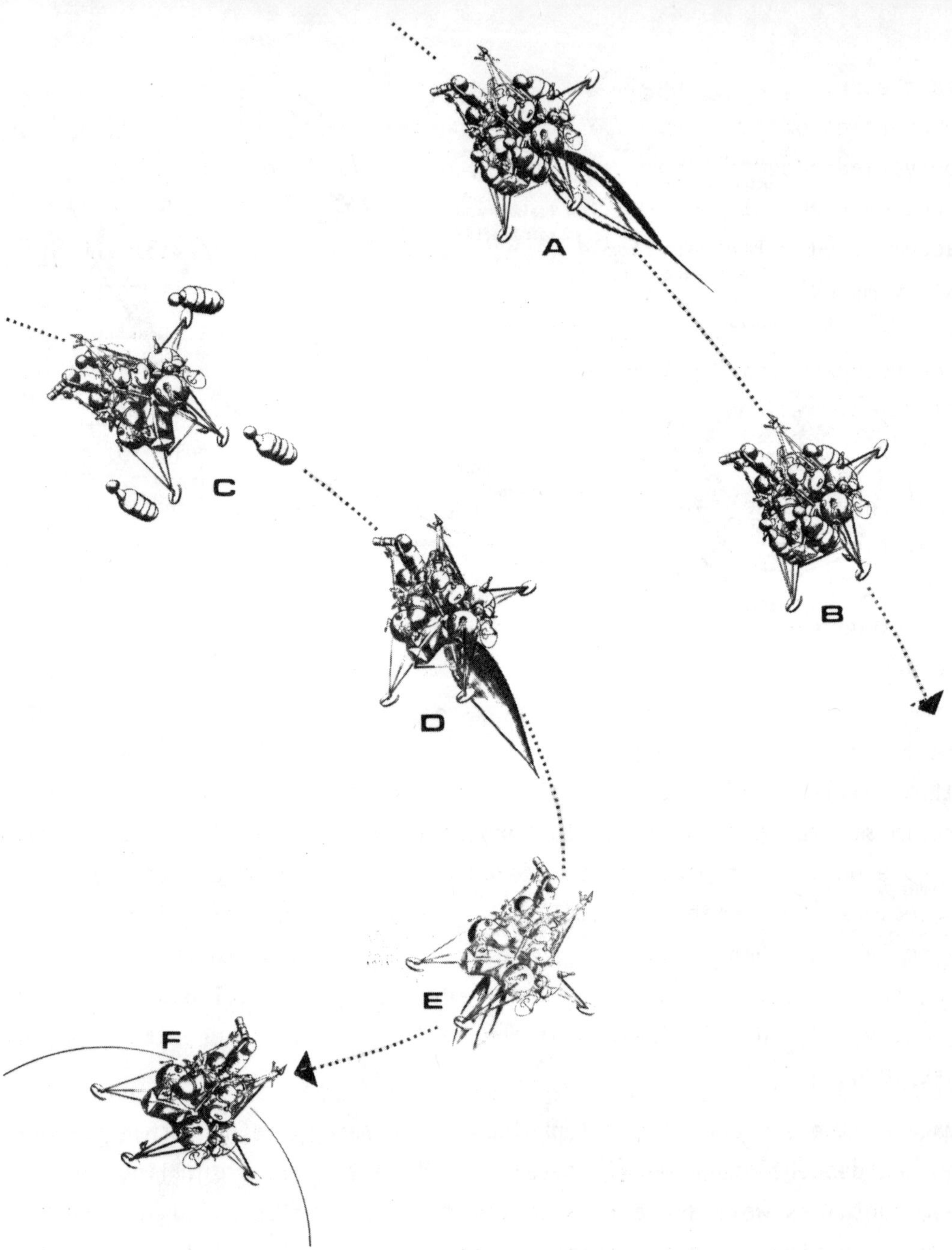

A. Lunar orbit injection burn
B. In lunar orbit
C. Ejection of depleted propellant tanks to descent
D. Main descent burn
E. Firing of smaller retro-rockets near the lunar surface
F. Lunar touchdown

Figure 28b Trajectory of Luna 16 to the moon (R. Escarcega)

to the drill, enabling Soviet scientists to determine characteristic properties of the lunar soil. The speed of rotation of the drill could be varied depending upon the density of the material being probed to prevent overheating. As the drill penetrated into the soil, the hollow bit accumulated a layered sample down to a depth of 35 cm. After the drilling was completed, the drill arm was raised, the drill head was rotated back 180 degrees, and the drill bit with the lunar sample was inserted into the return capsule. The drill head was then pulled back and an automatic device hermetically sealed the bit and sample within the return capsule (Figure 29). In all, the soil sample weighed approximately 101 gm.[85]

The ascent stage consisted of three major sections: the three spherical fuel tanks and engine nozzle, the cylindrical instrument compartment, and the 50-cm diameter spherical return capsule. The instrument section contained the necessary guidance equipment and chemical batteries. The return capsule, fastened to the instrument compartment by four metal straps (reminiscent of the Vostok and Venera spacecraft), was divided into three sections. The lower section contained batteries and transmitting equipment, the central section housed the lunar core sample and drill, and the upper section held the parachutes and the descent antennae.[86]

Twenty-six hours and twenty-five minutes after landing on the moon, the ascent-stage engine was fired on command from earth. After acquiring the needed 2.708 km/sec velocity the engine was shut down and the spacecraft began a three-day ballistic return to earth. Lunar launch and trans-earth insertion were so accurate that no course correction was needed on the return flight (Figure 30). At 0450 hr on 24 September 1970 approximately 48,000 km from earth the return capsule separated from the instrument section. The capsule entered the earth's atmosphere at 0810 hr while traveling at a velocity of about 11 km/sec. The craft was aerodynamically weighted so that the lower portion of the capsule covered with a heavy ablative material was pointed in the direction of travel. During reentry the temperature on the leading edge exceeded 10,000° C and the loads on the spacecraft reached 50 g's. Finally, the parachute compartment cover was ejected and a drogue chute was released at an altitude of 14.5 km and a velocity of 300 m/sec. The drogue chute was detached and the main

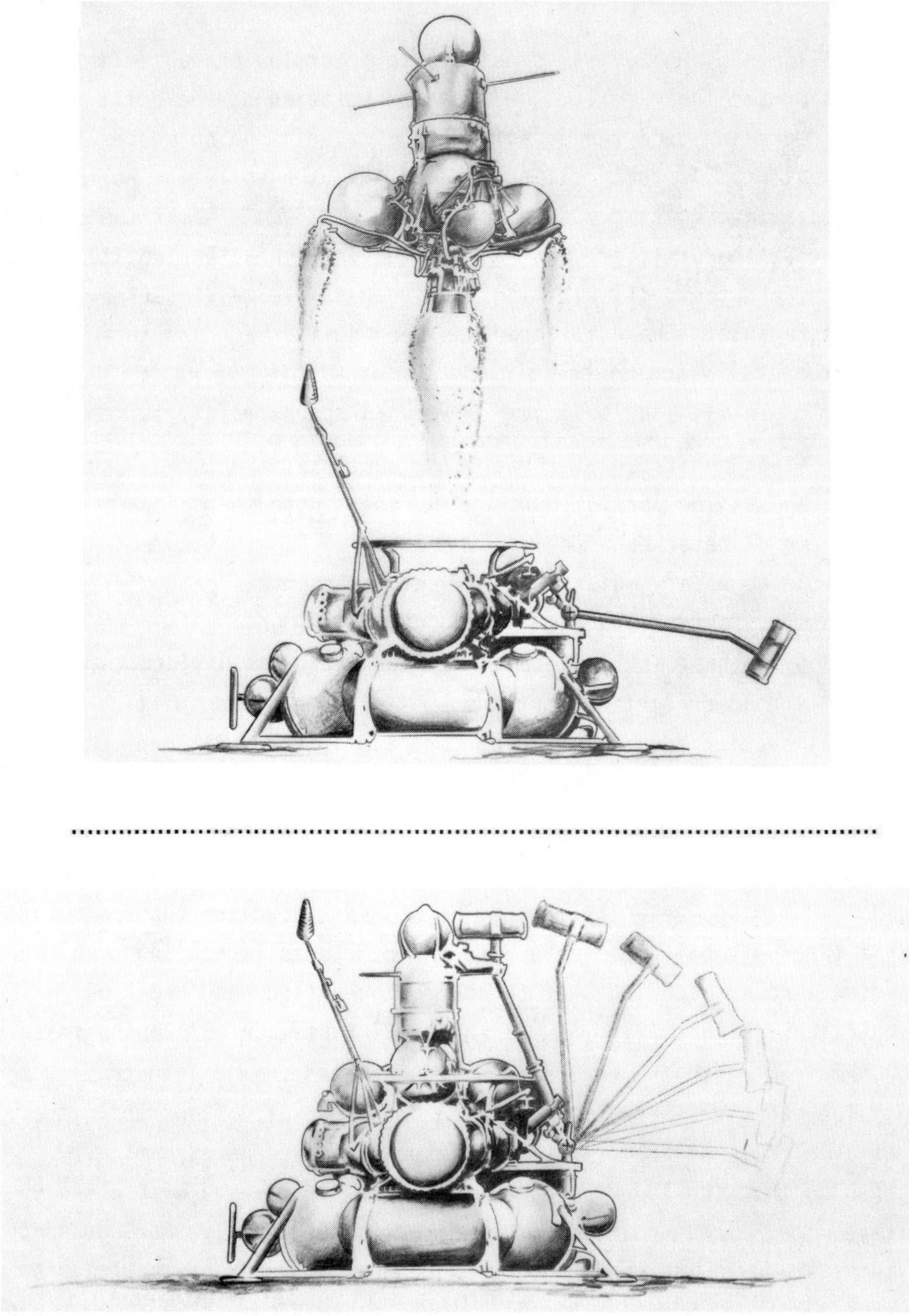

Figure 29 Luna 16's retrieval of a sample of lunar soil and blast-off from the moon (R. Escarcega)

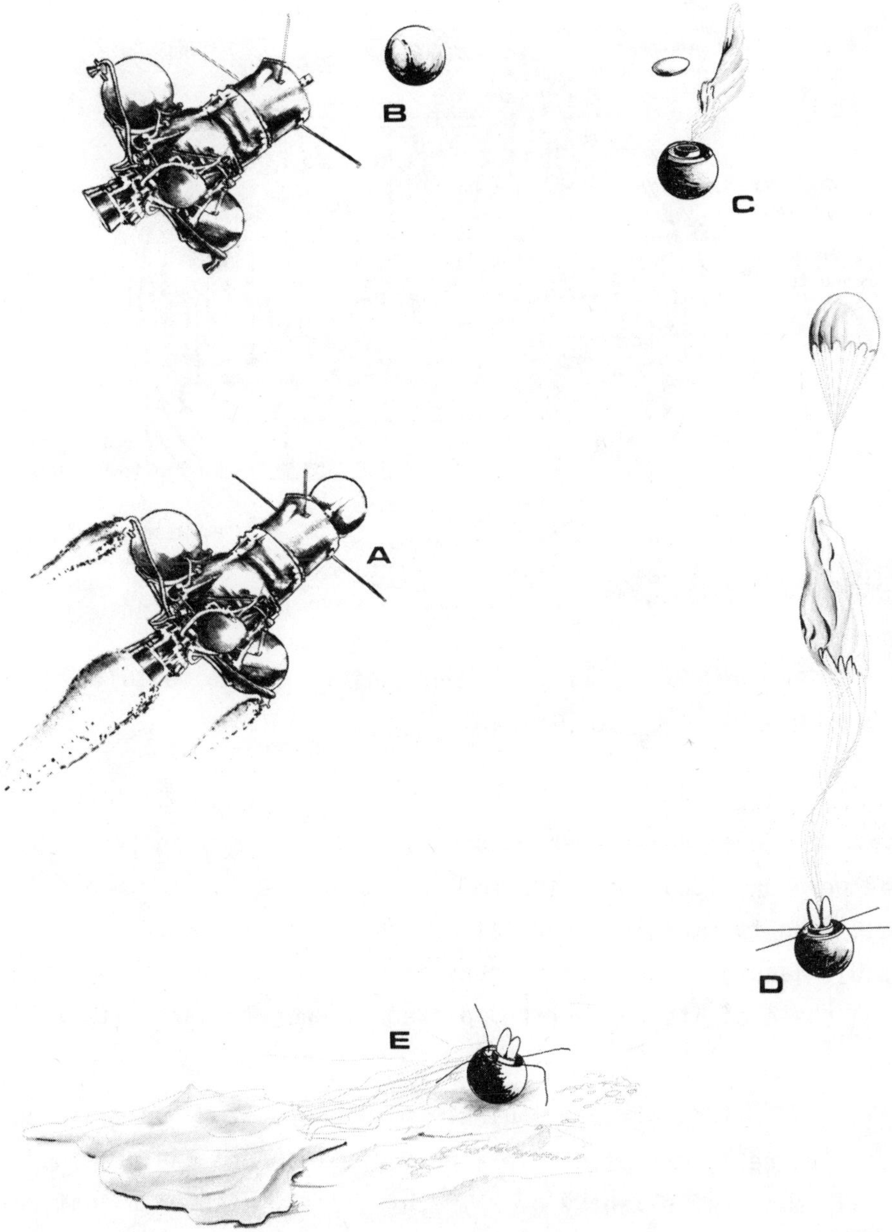

A. Blast-off from the moon
B. Ejection of the return capsule near earth
C. Drogue parachute deployment after reentry into earth's atmosphere
D. Deployment of the main parachute
E. Touchdown of the return capsule on Soviet territory

Figure 30a The return of Luna 16 to earth (R. Escarcega)

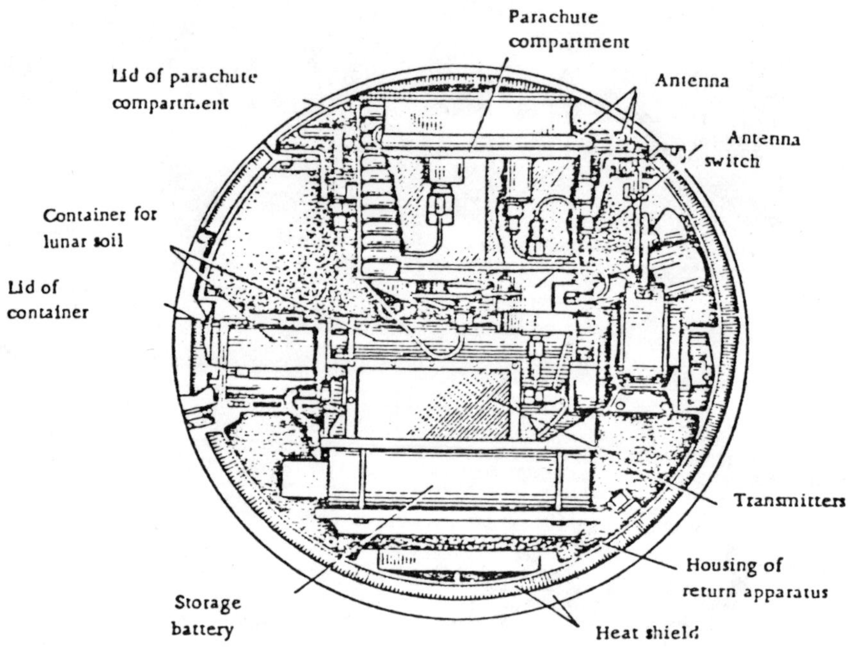

Figure 30b Diagram of the Luna 16 return capsule

parachute was deployed at an altitude of 11 km. Simultaneously, four whip antennae were unfolded and a transmitter activated to aid in locating the capsule. Also two long cylindrical balloons were inflated to ensure that the capsule landed in an upright position. The capsule was spotted during the final phase of its descent and picked up shortly after its 0826 hr landing.[87]

The drill and soil sample were quickly transported to a special receiving laboratory where it was placed in a helium atmosphere and examined. The lunar soil sample of slightly over 100 gm closely resembled that returned by the Apollo 11 and 12 astronauts, as evidenced by the accompanying table.[88] After suffering the remarkable successes of the US lunar programs the Soviet Union basked in the world praise of their unmanned lunar exploration. Less than two months were to pass before the USSR would demonstrate anew its technological prowess and its amazing ingenuity (Figure 31).

CHEMICAL COMPOSITION OF LUNAR ROCKS

	Basalt rocks of Luna 16	Fine fractions of Luna 16	Basalt rocks of Apollo 12	Fine fractions of Apollo 12
SiO_2	43.8	41.7	40	42
TiO_2	4.9	3.39	3.7	3.1
Al_2O_3	13.65	15.32	11.2	14
FeO	19.35	16.8	21.3	17
MgO	7.05	8.73	11.7	12
CaO	10.4	12.2	10.7	10
Na_2O	0.33	0.37	0.45	0.40
K_2O	0.15	0.10	0.065	0.18
MnO	0.2	0.21	0.26	0.25
Cr_2O_3	0.28	0.31	0.55	0.41
ZrO_2	0.04	0.015	0.023	0.09

Figure 31

LUNA 17

As early as 1955, two years before Sputnik 1 became the earth's first artificial satellite, Soviet scientists were contemplating a 500-kg tracked "tankette" equipped with television cameras for lunar exploration. Even fifteen years later after footprints had been left on the moon, the concept of an unmanned mobile lunar explorer controlled from earth was still largely considered an idea whose time had not yet come. Scarcely had extensive analysis of the Luna 16 soil sample begun when the launch of Luna 17 was announced.[89]

On 10 November 1970 at 1744 hr, Luna 17 was rocketed into an earth orbit of 237 km by 192 km at 51.6° by a Proton launch vehicle. From here the probe was quickly transferred to a lunar trajectory for the stated purpose of the "perfecting of new systems on board and the continuation of the scientific investigations of the moon and near-moon space." This simple announcement began a mission which was to last eleven months, far, far longer than any other Soviet lunar probe.[90]

Luna 17 was remarkably similar to Luna 16 in many ways. Weighing in the neighborhood of 5700 kg at launch, Luna 17 was divided into two sections: the descent stage which was virtually identical to that employed by Luna 16 and the mobile Lunokhod 1 vehicle. After landing on the moon, two pairs of ramps were extended from the descent stage and the Lunokhod vehicle perched on top was then able to descend in either direction (Figure 32).

The descent stage was basically unchanged from that of Luna 16, indicating the Soviet desire to standardize equipment whenever possible. In fact, this descent stage was subsequently used in all other third-generation Luna spacecraft including soil sample retrievers, Lunokhods, and heavy lunar orbiters. For the Lunokhods the descent stage instrumentation was slightly rearranged to allow for installation of ramps. In operation, the four large vertical cylindrical tanks were again jettisoned immediately prior to the deorbit burn. According to one Soviet report before the landing ramps were extended on to the lunar surface, the landing legs of the descent stage were blown away by pyrotechnic devices. The spacecraft was then resting with its empty fuel tanks on the surface of the moon. Apparently this procedure ensures that the spacecraft is level and the ramps are not deployed at too steep an angle. Since no other confirmation of this procedure has been found and photographic evidence indicates that Luna 17 and Luna 21 did not undergo such an operation, this procedure may represent an alternative which may or may not be used.[91]

The mobile Lunokhod 1 was an extremely sophisticated, self-contained explorer, although its appearance prompted observers to describe it as a bath tub on wheels with two large protruding eyes. The rover weighed approximately 756 kg, stood about 1.35 m high, and measured 2.15 m across the top of the instrument compartment (Figure 33). Mobility was achieved through eight independent electrically powered wheels 51 cm in diameter. With the wheels attached to Lunokhod 1 in four pairs, the distance from the front wheel to the rear wheel on each side was 2.216 m with 1.6 m between left and right side wheels. Each wheel was spoked and covered with a fine wire mesh for positive traction, allowing Lunokhod 1 to maneuver in two forward and reverse speeds. Turning was accomplished by either running the wheels on one side at a higher speed than the other

Figure 32 Final assembly of Luna 17/Lunokhod 1 prior to launch to the moon (TASS from SOVFOTO)

side or by actually setting one side of wheels in reverse with the other side in forward for sharper turns. Each wheel could be controlled separately and in the event of being caught or disabled any wheel could be uncoupled from the drive mechanism. Lunokhod 1 could maneuver with only two active wheels available.[92]

With speeds of up to 100 m/hr and an inherent time delay of about five seconds for signals to cover the round trip from moon to earth, several safety features were incorporated. Special sensing devices would automatically stop the vehicle if the inclination or list of the rover exceeded preset values, possibly preventing the Lunokhod from entering a crater from which it could not return. As already mentioned, Lunokhod 1 had the ability to travel in reverse, permitting the moon car not only to back out of precarious situations, but also to use either pair of

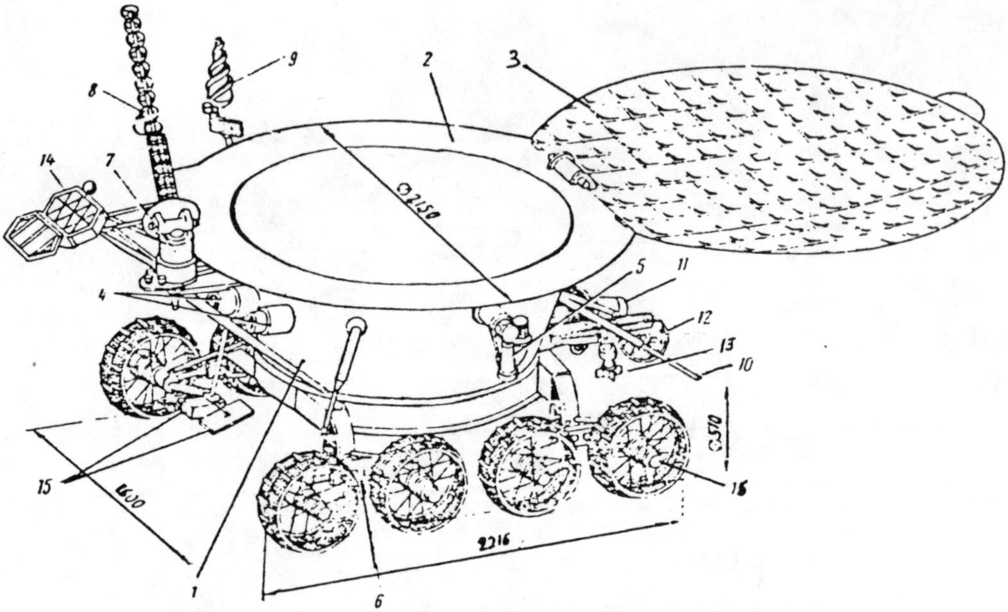

Diagram of the "Lunokhod-1" Self-Propelled Apparatus. 1, Body of the "Lunokhod" — hermetically sealed instrument section; 2, upper surface of the instrument compartment — radiator-cooler; 3, lid and solar battery panel; 4, television cameras; 5, telephotometers; 6, pair of wheels with torsion suspension; 7, drive mechanism for narrow-beam antenna; 8, narrow-beam antenna; 9, conical spherical broad beam antenna; 10, rod antenna; 11, isotopic source of thermal energy; 12, ninth wheel to measure distance traveled; 13, instrument for determining the permeability for ascertaining the physical and mechanical properties of the soil; 14, laser corner reflector; 15, remote unit of radio pulse measuring apparatus; 16, motor wheel.

Figure 33 Diagram of the Lunokhod 1 rover vehicle

ramps on the descent stage in the event that one was blocked. In order to determine how far the Lunokhod had traveled and establish its position relative to the descent stage, a ninth wheel was located at the rear of the vehicle to furnish information to an odometer as well as a speedometer. This ninth wheel could be raised or lowered on to the lunar surface whenever necessary. Each wheel also had a revolution counter which could be used as an odometer if necessary. However, the main wheels were more subject to slippage and thus distance measurements were not so accurate using this method. In fact, the traction properties of the lunar soil were estimated by measuring the slippage and torque on each wheel. The position of the Lunokhod was also furnished by pictures from the rover's TV cameras and by a system of gyroscopes which analyzed each turn the craft made.[93]

Of prime importance to earth controllers in guiding the Lunokhod across the moon was the television system. Two eye-like lens protruded from the front of the vehicle. Pictures taken through these cameras were transmitted every few seconds. Although not of high quality, they enabled the Lunokhod "driver" to steer around obstacles and maneuver close to interesting objects. When Lunokhod 1 was halted, a more detailed system of four photometers was employed. Two cameras were located on each side: one with a 360-degree vertical-plane scan and one with a 180-degree horizontal-plane scan. These cameras, quite similar to those used on Luna 9 and Luna 13, aided not only in navigation but also in panoramic analysis of the lunar surface. To obtain high-quality photographs a higher rate of transmission was needed, requiring the Lunokhod to stop and use its narrow-beam antenna.[94]

Lunokhod 1 was designed to do much more than merely ride along the lunar surface taking pictures. A large array of scientific experiments was attached to and within its magnesium-alloy instrument compartment. In order to determine soil composition an X-ray spectrometer similar in principle to the one used on Luna 13 was carried. This experiment, called RIFMA (Roentgen Isotopic Fluorescent Method of Analysis), relies on measuring the radioactivity of the soil elements after they have been irradiated by an isotopic source on board Lunokhod 1. The relative abundance of such elements as aluminum, calcium, silicon, iron, magnesium, and titanium could thus be calculated. In addition, the ten counters of the RIFMA instrument were able to detect the extent of radioactivity induced by solar flares.[95]

Further analysis of lunar soil mechanical properties was carried out by a penetrometer. Much refined over the projectile-type device on Luna 13, the Luna 17 penetrometer consisted of a blade studded cone and a mechanism for raising and lowering the cone on to the lunar surface. As the cone was pushed into the surface, the soil's resistance was measured. Then the cone was rotated and further resistance was recorded along with speed and angle of rotation.[96]

A French laser reflector experiment was mounted on the front of Lunokhod 1 just to one side of the two TV cameras. The 3.7-kg silica glass reflector consisted of 14 triangular prisms each 10 cm on a side in a housing approximately 46 cm by 20 cm. A cover was kept in place over the reflector during the lunar day and was opened when the vehicle had stopped for the lunar night. Both the Soviets and the French bounced 10^{-8} second ruby laser pulses off the reflector from earth observatories (Crimean Astrophysical Observatory of the USSR and Pic duMidi d'Ossau Observatory in France). By accurately measuring the time taken for the laser to reach the Lunokhod and be reflected back to the originating observatory, both the distance of the earth to the moon and the position of Lunokhod 1 on the moon could be calculated to a very high degree. The data could even be used to measure continental drift.[97]

Another valuable instrument carried by Lunokhod 1 was a radiometric detector designed to measure solar and cosmic radiation fluxes. One set of semi-conductor and gas-discharge counters was mounted on the exterior of Lunokhod 1 and could detect protons with energies above 1 Mev, electrons above 40 Kev, and alpha particles above 5 Mev. Inside the instrument compartment another gas-discharge counter could detect protons with energies above 30 Mev. These instruments were also capable of measuring any natural radioactivity from the moon itself.[98]

The last major experiment was an X-ray telescope. One X-ray photometer registered photons in the range of 2-10 Å, while a second counter was used to measure background radiation. Mounted along the lunar vertical the telescope was able to scan a 3-degree-wide path through space as the moon rotated. Each exposure lasted for six hours and could continue even when direct communications with earth were not possible. Due to the importance of mapping the X-ray universe from above the earth's atmosphere, this experiment was considered vital to feasibility plans for building large X-ray telescopes on the moon.[99]

To power all these experiments and the other equipment of Lunokhod 1, the Soviets used a combination of solar cells and chemical batteries. This was the first use of solar cells on a Luna probe since Luna 3 in 1959. The solar cells were attached to the underside of the lid of the Lunokhod. During the lunar day the vehicle would travel with the lid exposed, maintaining the charge of the storage batteries. During the lunar night when the vehicle was parked, the lid was closed to protect the solar cells and to provide additional insulation of the instrument compartment.

Due to the extreme temperature variations of the lunar day and night (in the neighborhood of +130°C and -150°C, respectively) and due to the temperature variations of objects in light or shadow during the lunar day (temperature on sunlit wheels could be +100°C while the temperature on wheels in shadow could be -30°C), an extensive thermal control system was needed. The thermal limits of the interior of the instrument compartment were 0°-40°C with an operating range of 15°-20°C desirable. During the lunar day the Lunokhod 1 vehicle was cooled by the forced circulation of gas through a heat exchanger in which water was evaporated. The excess heat was then radiated into space via a radiator in the top of the instrument compartment. At night the heart of the heating system was an isotopic heat source (probably polonium-210, see Luna 21). As the source underwent natural radioactive decay, helium nuclei were emitted which when trapped by the surrounding container generated enough thermal energy to heat the circulating gas via a heat exchanger and to maintain the compartment within permitted limits. Pressure within the instrument compartment was maintained within 730-780 mm Hg.[100]

Luna 17 completed the journey from the earth to the moon from 10 November to 15 November 1979 without incident, performing two course corrections on 12 and 14 November. In the very early morning hours of 15 November the Luna 17 spacecraft fired its main engine and entered a retrograde circular lunar orbit of 85 km at an inclination 141° with a period of 116 minutes. On 16 November an orbital maneuver dropped the perilune to 19 km. At 0641 hr on 17 November the main engine was again started and the descent to the lunar surface begun. Following a pattern similar to that described with Luna 16, two low-thrust engines were ignited just above the surface with touchdown occurring at 0647 hr with lunar coordinates of 38°17'N, 35°W, approximately 2500 km from the Luna 16 site (Figure 34).[101]

Shortly after landing, the television cameras were activated and the landing site was surveyed. The locking mechanism holding Lunokhod 1 on its platform was released and the hinged ramps were extended; then the Lunokhod vehicle edged slowly forward. After ensuring that the front two wheels did not sink into the lunar surface the entire rover rolled out on to the moon at 0928 hr (Figure 35). The next five earth days were cautious days of testing the Lunokhod's systems and learning its driving characteristics. Twenty meters were covered the first day and by 24 November a total distance of 197 m had been traveled.[102]

Between 24 November and 9 December Lunokhod 1 was parked and deactivated during the lunar night. The solar panel was closed and the French reflector cover opened with successful laser reflections occurring on 5 and 6 December. On 9 December Lunokhod 1 was reactivated and readied for its first full lunar day of exploration. The five-man Soviet teams directing Lunokhod 1 quickly gained experience and confidence. Driving sessions lasted longer and at higher speeds. Of course many stops were made to take soil analysis, photographs, and other readings. Once during this second day Lunokhod 1 automatically stopped while crossing a crater when the slope reached 27° and the list 17°. The lunar rover was shut down for three days during the lunar noon because of the inability to see accurately owing to bright reflections off the lunar surface and the lack of shadows. On 17 December Lunokhod 1 resumed operations and continued until 23 December when it was again parked to endure the lunar night after traveling 1.7 km.[103]

On 8 January 1971 the third lunar day began. One of the purposes of this third day of exploration was to head north and try to relocate the descent stage of Luna 17. Ten days later after a round trip of almost 4 km the Lunokhod vehicle returned to its starting point. Here it was to spend its third lunar night. A map of the first three lunar days of exploration is given in Figure 36. The Lunokhod vehicle was only designed for up to 90 days of operation. No one was sure whether the craft could survive a third lunar night. However, Lunokhod 1 was to amaze even its creators by eventually continuing to operate for eleven months until 4 October 1971 (although failure may actually have occurred in late September).[104]

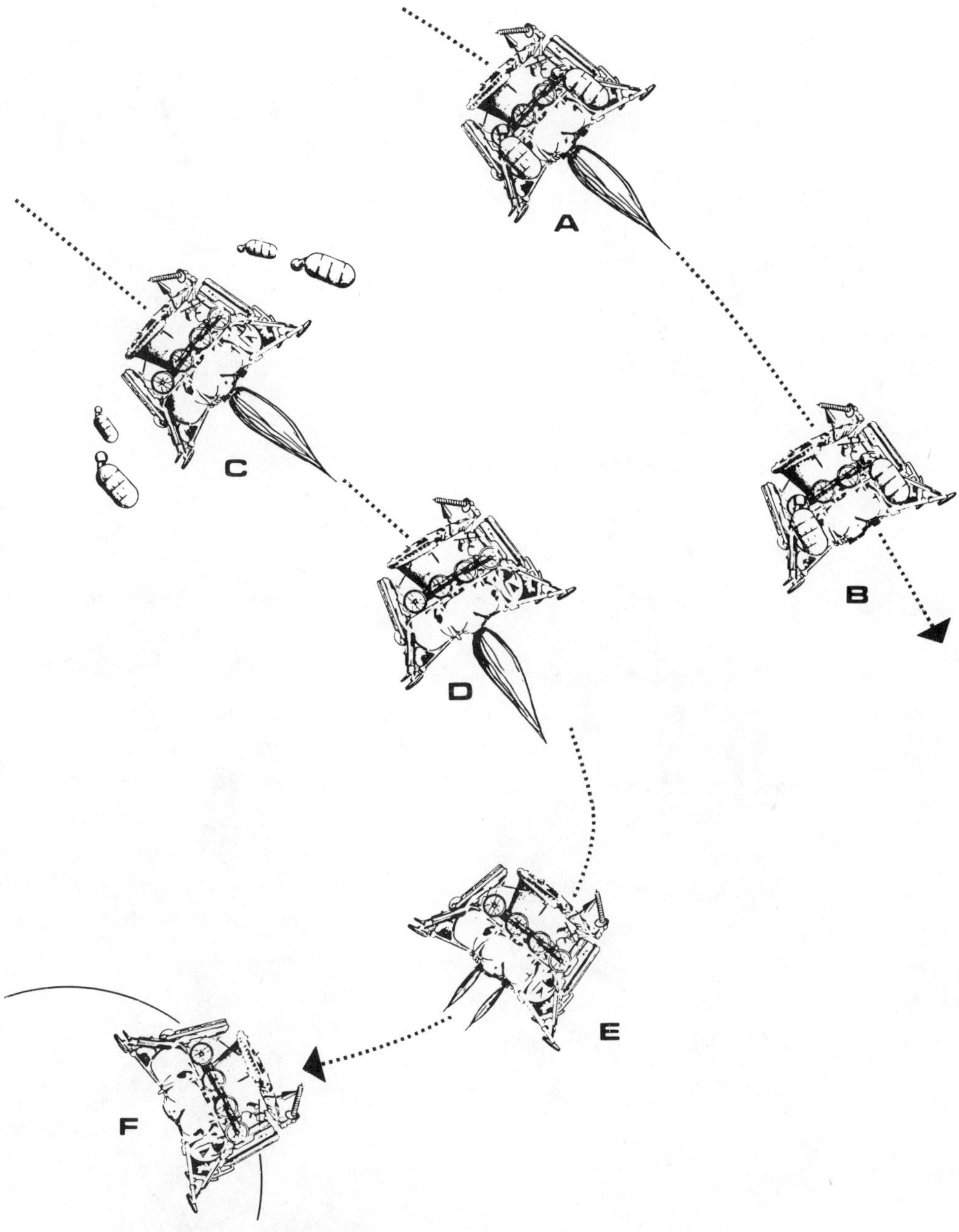

A. Lunar orbit insertion burn
B. In lunar orbit
C. Ejection of depleted propellant tanks prior to descent
D. Main descent burn
E. Firing of smaller retro-rockets near the lunar surface
F. Lunar touchdown

Figure 34 Luna 17 trajectory to the moon and landing sequence (R. Escarcega)

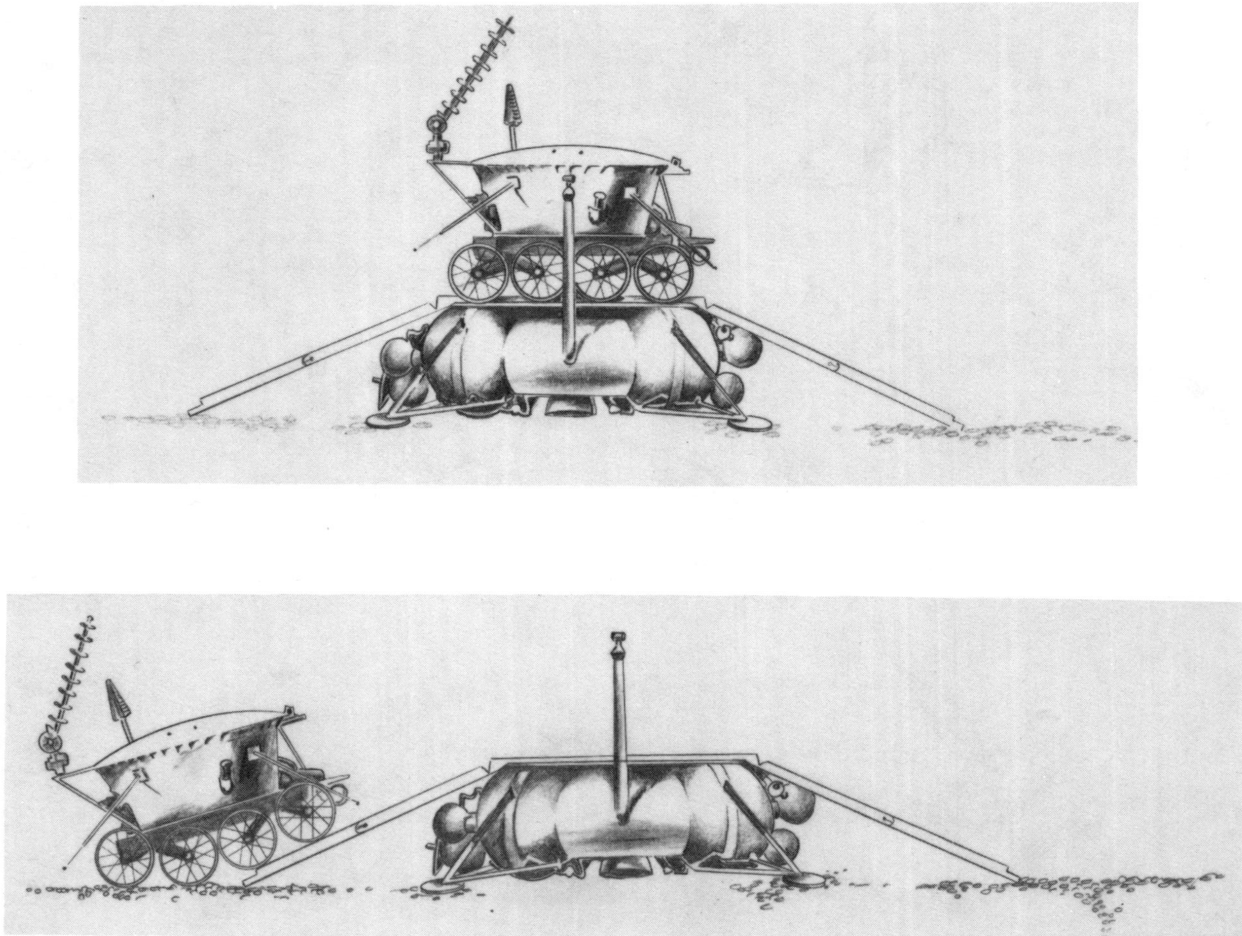

Figure 35 The descent of Lunokhod 1 onto the lunar surface from the Luna 17 descent stage (R. Escarcega)

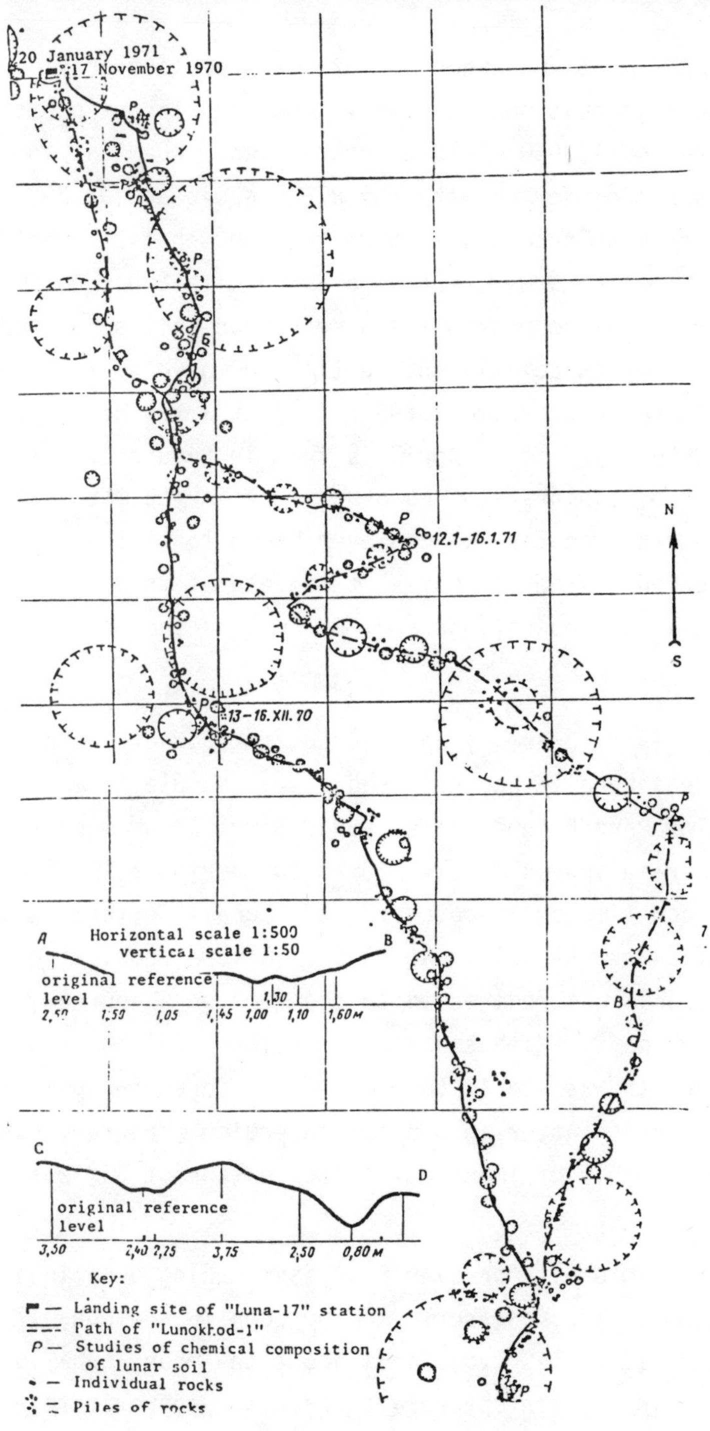

Figure 36 Map of the route of Lunokhod 1 on the surface of the moon from 17 November 1970 to 19 January 1971

Month after lunar month Lunokhod 1 performed spectacularly. By the time Lunokhod's isotopic heat source finally could no longer protect the craft, a total distance of 10,540 m had been traveled. Chemical analysis of lunar soil and rocks was performed with the RIFMA equipment at 25 locations, while the penetrometer took readings at more than 500 spots. Twenty-five thousand television pictures were returned along with 200 panoramas. On 10 February 1971 Lunokhod encountered and observed an eclipse of the sun by the earth. This was an especially interesting engineering test since temperatures on the moon dropped almost 290°C in just a few hours. A map of its travels from 7 February to 13 April is seen in Figure 37. Throughout its entire 322 days of exploration, Lunokhod 1 performed more experiments and conducted more research than we can even begin to list here. The interested reader is referred to the footnotes at the end of this section for more details of the actual operations and for some of the scientific findings.[105]

LUNA 18

As Lunokhod 1's last lunar day approached, the Soviets were busy readying Luna 18. Whereas Luna 16 had returned a soil sample from the relatively smooth plains of Mare Fecunditatis, Luna 18 was programmed to land in mountainous terrain approximately 120 km due north of the Luna 16 site. At 1641 hr Moscow time on 2 September 1971 Luna 18 entered a temporary earth orbit of 242 km by 186 km at 51.6°. Before completing an entire revolution, Luna 18 was redirected to the moon on a journey which would take almost five days. With course corrections on 4 and 6 September, by 7 September Luna 18 was reoriented with its main engine pointing in the direction of flight. After an orbital insertion maneuver, Luna 18 entered a circular lunar orbit of 100km at an inclination of 35° with a period of 119 minutes.[106]

Luna 18 remained in orbit for over four days making occasional orbital corrections similar to those performed by Luna 16 and Luna 17. Finally, on 11 September, after 54 revolutions about the moon, Luna 18's retrorocket fired again and the spacecraft began the approximately five-minute descent. Unfortunately, all communications were lost with Luna 18 at 1048 hr at the calculated moment of impact. Luna 18 had apparently crashed in rough terrain at 3°34'N, 56°30'E. A Soviet announcement simply stated that the Luna 18 landing "under the complex conditions of the lunar continent was unfavorable."[107]

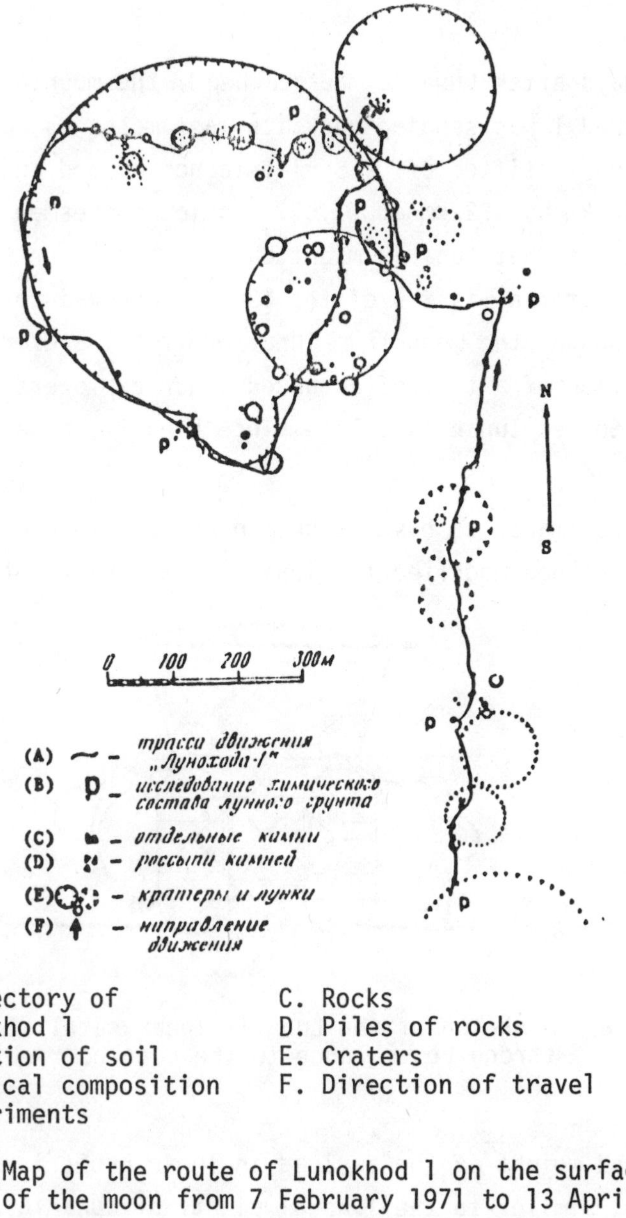

A. Trajectory of Lunokhod 1
B. Location of soil chemical composition experiments
C. Rocks
D. Piles of rocks
E. Craters
F. Direction of travel

Figure 37 Map of the route of Lunokhod 1 on the surface of the moon from 7 February 1971 to 13 April 1971

Due to the premature end of the Luna 18 flight, little scientific data was obtained. However, during the Eighteenth COSPAR meeting in Varna, Bulgaria, in 1975 Soviet scientists revealed that analysis of the Luna 18 continuous-wave radioaltimeter (3.1 cm wavelength) data had been completed. From the readings returned it was possible to determine the dielectric constant of the lunar surface down to a depth of 0.5 m which in turn indicated a mean density of the top lunar soil of 0.8 ± 1.5 g/cm^3.[108]

LUNA 19

Seventeen days after Luna 18 had crashed in the mountains of the moon and while Lunokhod 1 was sending back its last messages, a Proton rocket carrying Luna 19 lifted off a Soviet launching pad and soared into an earth orbit of 260 km by 172 km at 51.6°. Soviet spokesmen quickly dispelled any suggestions that Luna 19 might carry a sample retriever or Lunokhod by stating that the mission of the lunar probe was to conduct scientific investigations of the moon from lunar orbit. Therefore Luna 19 was the third variation of a trio of advanced lunar explorers and would be the first real Soviet lunar satellite since Luna 14 three and one-half years earlier.[109]

Luna 19 is estimated to have weighed near 5900 kg and may have carried a wheelless Lunokhod modified for lunar orbit experiments (Figure 38).

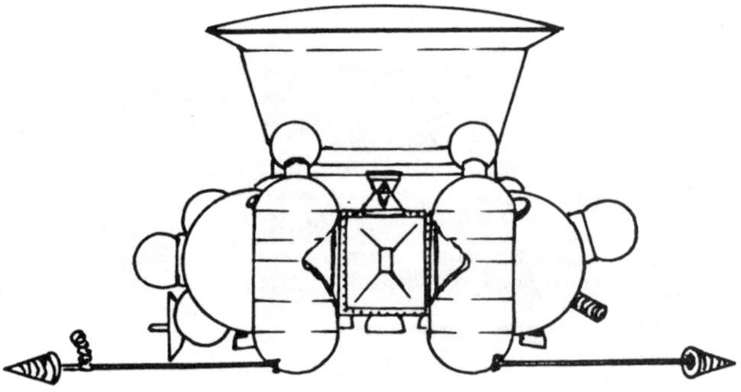

Figure 38 A sketch of the Luna 12 lunar satellite. Note the strong resemblance to the Lunokhod vehicle (R.F. Gibbons)

The estimated weight of the payload in lunar orbit is between 1850 kg and 1950 kg, similar to the landed weights of Luna 16 and Luna 17. After course corrections on 29 September and 1 October Luna 19 slipped into a 140-km circular lunar orbit with inclination of 40°35' on 3 October 1971, following an uneventful flight from the earth to the moon. Its period of revolution was 2 hours 1 minute and 45 seconds. By 7 October the orbit had been altered to 135 km by 127 km. Later in its mission Luna 19 moved to a 385 km by 77 km orbit.[110]

Luna 19 was equipped to conduct many scientific experiments. One mission of the spacecraft was to provide panoramas of the mountainous region of the moon between 30° and 60° S lattitude and between 20° and 80° E longitude with its image-scanning television cameras. Another experiment involved the use of a gamma-ray spectrometer to determine the surface chemical composition. Using the readings of the radioaltimeter for the region near 41°N, 54°E such surface properties as density, dielectric constant, and porosity were calculated. Electron concentrations between altitudes of 5 and 35 km above the moon's surface were also measured. Micrometeor detectors registered the direction, mass, and energy of such particles. Magnetometers on board Luna 19 examined the disturbance caused by the moon on the earth's magnetosphere and on the interplanetary magnetic field. Finally, aided by the long lifetime of Luna 19, extensive studies were performed on the shape and strength of the lunar gravitational field and the location of mascons.[111]

Sometime between 3 and 20 October 1972 the Luna 19 mission was completed when, after over one year and 4000 revolutions, Luna 19 became silent. Unlike US practices, the USSR apparently does not crash its lunar satellites into the lunar surface upon completion of their missions. This procedure by the US with Lunar Orbiter spacecraft was designed to prevent any possible interference with later spacecraft.[112]

LUNA 20

The twentieth acknowledged probe of the Luna series was successfully launched on 14 February 1972 with the purpose of completing the mission of Luna 18: the return of soil from the lunar continents. After leaving an earth orbit of 238 km by 191 km at 51.5°, Luna 20 began the four-and-one-half-day coast to the moon. A mid-course correction maneuver was performed on 15 February. By 18 February Luna 20 had become a temporary satellite of the moon in a circular orbit of 100 km with an inclination of 65°. On 19 February the perilune was lowered to 21 km in preparation for a landing attempt.[113]

At 2213 hr Moscow time on 21 February, Luna 20 responded to a command from earth and fired her main engine for 267 seconds. With this deceleration impulse the spacecraft rapidly dropped from orbit. At an altitude of

760 m (160 m higher than Luna 16) the main engine was restarted to further reduce the velocity of the probe. When Luna 20 was only 20 m above the surface, the main engine was shut down and two smaller engines were ignited. After a descent of only six minutes Luna 20 landed at 2219 hr. The landing site of 3°32'N, 56°33'E was only about 1.8 km from the crash site of Luna 18 and 120 km from the Luna 16 site.[114]

Luna 20 had landed during the lunar day in contrast to Luna 16 which had landed at night. Almost immediately after landing, Soviet scientists activated the craft's television cameras and searched for the most promising spot from which to extract a sample. After an interesting drilling site had been chosen, the drilling arm and rotary drill were unlocked from their stored position and the drill head was lowered to the lunar surface--a procedure which lasted approximately twelve minutes. The drill closely resembled that used by Luna 16. Capable of speeds up to 500 rpm the rotary drill hit hard rock at a depth of 100-150 mm. The drill continued its penetration, but due to the tough drilling conditions it was automatically shut down three times by an overload protection circuit to prevent overheating. After the drilling was complete, the sample was withdrawn and lifted up to the return capsule atop Luna 20. The lunar soil and rock were inserted and pyrotechnic devices hermetically sealed the capsule.[115]

Luna 20 remained on the lunar surface with its valuable payload until 0158 hr on 23 February. At that time the ascent stage lifted off the moon and quickly accelerated to the 2.7 km/sec required to return to earth. On 25 February at a distance of almost 52,000 km the descent capsule was detached from its carrier and entered the earth's atmosphere at an angle of 30° to the horizontal. Following the deployment of the drogue and main parachutes, the Luna 20 landing capsule finally came to rest at 2219 hr on an island in the Karkingir River 40 kilometers north of Dzhezkagan (48°N, 67°34'E) during a snow storm. The capsule was soon located and recovered and transported to the USSR receiving laboratory.[116]

The lunar soil and rock were immediately removed, and the long process of analysis began. There is some uncertainty of the depth at which the sample was taken and of the weight of the sample. Most reports simply assume that the depth and weight figures given for Luna 16 apply also to Luna 20. Soviet spokesmen and publications have usually ignored the question. As late as 1976 figures of 50-100 grams for the weight and up to 33 cm for the depth have been reported. It is possible that since the drill did encounter some hard rock, a full sample was not obtained.[117]

The rocks and soil were submitted to a variety of tests in which over seventy elements were identified. The sample differed from that of Luna 16 in that the majority of the rock particles of Luna 20 were described as anorthosite rather than the basalt of Luna 16. Many of the findings of these tests can be found in the references at the end of this section. In addition, the Luna 20 radioaltimeter readings were used to determine the lunar soil density, etc, in the same manner as Luna 19.[118]

LUNA 21

Less than a month after the final US manned moon landing, the USSR renewed its lunar exploration program with the launch of Luna 21. On 8 January 1973, Luna 21 entered an earth orbit of 236 km by 183 km at 51.6°. Within little more than an hour, Luna 21 along with the Lunokhod 2 which it carried was streaking toward the moon. A course correction was performed the next day and on 12 January the spacecraft reached the vicinity of the moon. During this period the lid of the Lunokhod 2 rover was opened to expose its solar cells and charge its internal batteries. While undergoing spacecraft maneuvers, the lid was closed.[119]

After entering a lunar orbit of 100 km by 90 km with an inclination of 60° on 12 January, Luna 21 remained in orbit for almost four days during which time the perilune was lowered to 16 km. On 16 January Luna 21's main engine was fired and the spacecraft dropped from orbit. At an altitude of 3 km the engine shut down and Luna 21 fell toward the moon. The main engine was burned again from 735 m down to 20 m where the two smaller engines took over and further reduced the spacecraft velocity to only 2 m/sec. Luna 21 came to rest at 0135 hr on 16 January inside the crater LeMonnier at 25°51'N, 30°27'E. Luna 21 was thus able to examine a variety of terrain and soil conditions between the Mare Serenitatis and the Taurus Mountains.[120]

The 1.35-m tall Lunokhod 2 was an improved version of Lunokhod 1. Lunokhod 2 weighed 840 kg, 84 kg more than Lunokhod 1. One addition was the placement of a third television camera in front of the vehicle at a level even with the top of the instrument compartment, permitting an extended view for the earthbound controllers. The television system had also been refined to return new photos of higher quality every three seconds. The eight-wheel traction system had been improved, providing Lunokhod 2 with the capability of traveling twice as fast as its predecessor and making turns while in motion. The French supplied another laser reflector, and the RIFMA experiment for determining the chemical composition of the lunar soil had been improved. The Soviets also now revealed that the isotopic heat source was polonium-210.[121]

Three major new experiments had also been added. An astrophotometer was installed to measure visible and ultraviolet light levels. This device was designed to measure any luminosity of the lunar sky which might hinder the operation of a future lunar observatory and to search out and analyze the zodiacal light. Similar devices had been carried on board the Kosmos 51 and 213 earth satellites. The second new experiment was a magnetometer located at the end of a 2.5-m boom in front of the vehicle. As well as measuring any overall magnetic field, it could detect specific areas of anomalous magnetism. The third experiment used a photodetector called Rubin-1 to aid in laser detection experiments.[122]

After a brief survey of the landing site after touchdown, Lunokhod 2 was released and rolled out on to the surface at 0414 hr on 16 January. The rover was then moved a short distance from the lander and allowed to charge its batteries for the next two days. Lunokhod 1 had not required this extra time for charging, which lent support to reports that Luna 21 had experienced some power supply problems. Although off to a slow start, Lunokhod 2 quickly redeemed itself. Day after day the lunar rover rolled up and down steep slopes and through craters as Soviet drivers taxed Lunokhod 2 to its limits. Several times the moon car attempted crater walls too steep and once the wheels sank half way into lunar dust. Each time, however, the rover was able to back up or go around and continue.[123]

By the end of the second lunar day on 23 February Lunokhod had already traveled farther than Lunokhod 1 had in its entire year of active life.

Lunokhod 2 continued to explore and conduct experiments for five lunar days before the USSR announced the shutdown of Lunokhod 2 had taken place on 3 June 1973, although evidence points to a possible breakdown in mid-May. Apparently the failure of Lunokhod 2 happened suddenly since the vehicle was not left in a position in which the French laser reflector could still be utilized. Although Lunokhod 2 had lasted less than one-half as long as Lunokhod 1, the improved rover traveled more than 3.5 times as far. Eighty thousand television pictures and 86 panoramic photographs were returned, and hundreds of mechanical and chemical soil tests were performed. Readings of the lunar sky during sunlit hours revealed an unexpected glow would hinder lunar observatory work. The data from these and other experiments would keep Soviet scientists busy for years to come.[124]

LUNA 22

The third-generation Luna spacecraft were characterized as part of a well-balanced exploration program. With the completion of the Luna 21 mission, the Soviets had flown two successful sample return missions (16 and 20), two successful lunar rover missions (17 and 21), and one successful heavy lunar satellite (19). On 29 May 1974, one year after Lunokhod 2 had ceased functioning, the Soviets evened the score with the launch of Luna 22 into an earth orbit of 227 km by 178 km at 51.6°. Shortly thereafter the USSR's second heavy lunar satellite was on its way to a rendezvous with the moon. After a course correction on 30 May, Luna 22 coasted without incident to the vicinity of the moon. As the probe neared the moon, its main engine fired and Luna 22 slipped into a circular 220 km orbit inclined 19°35' to the lunar equator.[125]

Luna 22 is believed to have been virtually identical to Luna 19: its primary mission a photographic survey of the moon. However, many other experiments were performed, including determining chemical composition of the lunar surface through radioaltimeter readings, registering meteor activity in lunar orbit, searching for a lunar magnetic field, measuring solar and cosmic radiation flux, and further investigating the irregular lunar gravitational field. The presence of a lunar atmosphere and electron concentrations was studied by the use of a two-frequency dispersion interferometer. To perform the television survey, Luna 22's orbit was altered on 9 June to 244 km by 25 km. By 13 June the survey had been completed and Luna 22 was placed in a new orbit of 299 km by 181 km to begin the gravitational field studies. Between June and August 1974 Luna 22 counted 23 impacts from meteors of mass greater than 10^{-11} gm. This was a decrease of a factor of three compared with the meteor flux at the lower orbits in the beginning of June.[126]

On 11 November Luna 22 was again shifted to probe circumlunar space at higher altitudes. Its new orbit was 1437 km by 171 km at an inclination of 19°33' where Luna 22 now required three hours and 12 minutes to complete one revolution of the moon. By 2 April 1975 the orbit of Luna 22 had been perturbed to 1409 km by 200 km at 21°.[127]

On 2 June 1975 Luna 22 had been in lunar orbit for one year, and the satellite was still functioning after 3296 orbits. After almost fifteen months in orbit Luna 22's perilune was dropped to only 30 km on 24 August 1975. In an apparent test of total system equipment, the television system was again activated and a photograph returned to earth. This represents one of the longest periods of any Soviet deep-space television system to function. At the conclusion of the test, Luna 22 was placed in an orbit of 1286 km by 100 km at the same 21° inclination. After exhausting its maneuvering fuel on 2 September, Luna 22 was reported to have ended its mission in early November 1975.[128]

LUNA 23

While Luna 22 was only one-third the way through its mission, the USSR was readying yet another probe of the moon. A Proton launch vehicle lifted Luna 23 into an orbit of 246 km by 182 km with inclination of 51.5° on 28 October 1974. Luna 23's mission was to return a deep core sample from beneath the surface of the moon. The maximum depth of the Luna 16 and Luna 20 drilling apparatus was approximately 0.3 m. Luna 23, however, carried an improved drilling head capable of extracting samples from a depth of 2.5 m.[129]

After a trajectory correction on 31 October, Luna 23 fired its main engine on 2 November and braked into a lunar orbit of 104 km by 94 km at an inclination of 138°. The orbit was altered on 4 and 5 November to 105 km by 17 km. On 6 November Luna 23 was readied for the descent burn. Following the same procedure as the previous third-generation landers, Luna 23 decelerated from orbit and came to rest in the southernmost part of Mare Crisium. Unfortunately, the Soviets reported that during landing in unfavorable terrain the drilling device had been damaged and was rendered inoperable for its primary mission.[130]

Unable to obtain any lunar samples, Soviet scientists reportedly decided that a return flight was unnecessary. "A newly formulated abbreviated program, including the testing of new construction parts and equipment of automatic lunar stations" was then implemented. Contact with the probe was maintained until 9 November 1974.[131]

LUNA 24

Almost two years passed since the launching of Luna 23 before the next and, thus far, final probe of the Luna series. Launched into an earth orbit of 243 km by 188 km at 51.5° on 9 August 1976, Luna 24 was programmed to complete the mission Luna 23 had failed to accomplish. A second attempt was being made due to the high priority placed on returning a sample from Mare Crisium, site of a large lunar mascon. A routine course correction was performed on 11 August, and at 0211 hr on 14 August Luna 24 was decelerated and captured by the moon in a 115-km circular orbit inclined 120° to the lunar equator.[132]

During 16 and 17 August the orbit was modified to 120 km by 12 km. Luna 24's main engine was fired at 0930 hr on 18 August, and six minutes later Luna 24 had come to rest unharmed at 12°45'N, 62°12'E, virtually the same location as the less fortunate Luna 23 and not far from the spot where the mysterious Luna 15 had crashed. This time, however, the drilling apparatus was intact. After carefully unlatching and lowering the drilling head over the selected site, Soviet controllers activated the rotary percussion drill. Apparently without difficulty a core sample was obtained to a depth of 2 m, followed by storage of the sample in 30-cm segments within the return capsule. Finally the return capsule access hatch was sealed, and Luna 24 awaited instructions for return.[133]

Less than 23 hours after landing on the moon, the ascent stage lifted off at 0825 hr on 19 August and headed home. A Soviet announcement 24 August reported communications were still being held with the descent stage left on the moon. The return capsule was separated from its carrier rocket and entered on the earth's atmosphere on the morning of 23 August. At an altitude of 15 km the parachutes and transmitting antennae were deployed, and the capsule with its payload landed at 0855 hr, 200 km SE of Surgut. Two days later the first reports on the soil and rock strata indicated a laminated type structure, characteristic of a layering process on the moon. Although not confirmed, the weight of the Luna 24 sample is believed to be between 140 and several hundred grams with the former weight the most probable.[134]

FUTURE MISSIONS

Over three years have passed since the return of Luna 24. The USSR has demonstrated that she will continue to pursue a steady, inventive program of lunar exploration. What might one expect in the future? Although further use of spacecraft of the Luna 16 and Luna 17 type is possible, two new mission profiles have been outlined. The first and probably the simpler is a far-side retriever. A Luna 16 type spacecraft would land on the far side of the moon, scoop up a soil sample, and return to earth. This mission would probably require a lunar satellite to relay communications to and from earth and the spacecraft while on the far side. Whereas it might be possible to program the spacecraft to perform its mission without guidance from earth, the problem of accurately fixing its position to determine time and flight path for lift-off is much harder. The Soviet Union has already announced plans for a lunar polar orbiter in the 1980's which may be related to the far-side retriever effort.[135]

The second mission profile probably under consideration would be a combination rover-retriever. With mobility the spacecraft could travel to many different locations to collect the most interesting and diverse samples. Then, when all samples had been collected, the probe could launch itself from the moon and return to earth. The weight of such a spacecraft appears to be beyond the capacity of the Proton launch vehicle, but within the reach of the oft-mentioned G-class launch vehicle. One method of overcoming this obstacle, however, is dual launches of D-class rockets. Under this plan both a Luna 16 and a Luna 17 type spacecraft would be landed in in the same general region of the moon. The Lunokhod could collect the samples and then deliver them to the retriever which would store the payload and blast off for earth.[136]

REFERENCES AND NOTES

1. There are unconfirmed reports of several lunar attempts in 1958. Most notably on 1 May, 24 Sep., and 26 Nov. See *Spaceflight*, July-Aug. 1977 p. 276 and Robert Beard's *Soviet Space Failures*, 1970.

2. Skuridin, G.A., *Mastery of Outer Space in the USSR, 1957-1967*, pp. 310-327; Wukelic, G.E., *Handbook of Soviet Space-Science Research*, p. 62

3. Skuriden, G.A., *op. cit.*, pp. 332-334

4. Blagonravov, A.A., et al., *USSR Achievements in Space Research*, pp. 219-220; Skuridin, G.A., *Op. cit.*, p. 168; Aleksandrov, S.G. and Federov, R. Ye., *Soviet Satellites and Space Ships*, pp. 149-151

5. Wukelic, G.E., *op. cit.*, pp. 83, 86-87; Skuridin, G.A., *op. cit.*, pp. 348-351

6. Wukelic, G.E., *op. cit.*, pp. 244, 260, 274, 284; Blagonravov, A.A., et al., *op. cit.*, pp. 119-120; Skuridin, G.A., *op. cit.*, pp. 340-348

7. Petrovich, G.V., *The Soviet Encyclopedia of Space Flight*, pp. 39-40; Zhdanov, G. and Tindo, I., *Space Laboratories*, pp. 39-40; Skuridin, G.A., *op. cit.*, p. 594; Caidin, M., *Red Star in Space*, pp. 166-171; Gatland, K., *Robot Explorers*, p. 130

8. *Soviet Space Programs, 1971-1975*, p. 84

9. For a brief summary of early U.S. lunar program see Gatland's *Robot Explorers* and Ley's *Ranger to the Moon*.

10. *Soviet Moon Rockets*, p. 8; Skuridin, G.A., *op. cit.*, pp. 365-368; *Nature*, 15 Aug. 1959, p. 502 and 2 Jan. 1960, p. 11

11. *Ten Years of Space Exploration*, p. 19; *USSR Explores the Moon*, p. 3; Blagonravov, A.A., et al., *op. cit.*, pp. 278, 596-598; Skuridin, G.A., *op. cit.*, pp. 358, 366-367; Wukelic, G.E., *op. cit.*, pp. 83, 87, 168, 224, 243, 244, 260-261, 275-276, 284; Aleksandrov, S.G. and Federov, R. Ye., *op. cit*, pp. 80-84

12. Petrovich, G.V., *op. cit.*, p. 76; Skuridin, G.A., *op. cit.*, pp. 375, 365, 379; Blagonravov, A.A., *op. cit.*, pp. 488, 598; *Soviet Moon Rockets*, p. 24

13. *Space Laboratories*, pp. 42-44

14. Zaehringer, A.J., *Soviet Space Technology*, p. 143

15. Gatland, K., *op. cit.*, pp. 131-132. Skuridin, G.A., *op. cit.*, pp. 382, 390-391; *Space Laboratories*, pp. 44-46; Wukelic, G.E., *op. cit.*, pp. 83, 87, 224, 261, 276-277, 285

16. *An Atlas of the Moon's Far Side; Ten Years of Space Exploration*, p. 19; Petrovich, G.V., *op. cit.*, pp. 232-233; Blagonravov, A.A., et al., *op. cit.*, pp. 279-282, 598; Skuridin, G.A., *op. cit.*, pp. 391-392; *USSR Explores the Moon*, pp. 4-5

17. Riabchikov, Evgeny, *Russians in Space*, p. 154; Shelton, W., *Soviet Space Exploration: The First Decade*, p. 77; Skuridin, G.A., *op. cit.*, pp. 131-132; *Soviet Moon Rockets*, pp. 40-52; *USSR Explores the Moon*, pp. 4-5. (For more details of Luna 3 see *Atlas of the Reverse Side of the Moon, Part 2; An Atlas of the Moon's Far Side;* and *Soviet Satellites and Spaceships.)*

18. Skuridin, G.A., *op. cit.*, p. 410; Shelton, W., *op. cit.*, p. 77; *Soviet Space Programs, 1966-1970*, p. 163; *Spaceflight*, Nov. 1969, p. 378; Ley, W., *op. cit.*, pp. 98-99; Rebrov, M. and Khozin, G., *The Moon and Man*, p. 99; Blagonravov, A.A., et al., *op. cit.*, p. 599

19. Caidin, M., *op. cit.*, pp. 180-181; *TRW Space Log, 1975*, p. 39; *Soviet Space Programs*, 1971-1975, p. 554

20. *Astronautics and Aeronautics*, May 1963, p. 134; *Spaceflight*, July-Aug. 1977, p. 276; Gatland, K., *op. cit.*, p. 133; *TRW Space Log, 1975*, p. 45; *Soviet Space Programs, 1971-1975*, pp. 93, 556

21. Gatland, K., *op. cit.*, p. 133; *Astronautics and Aeronautics*, May 1963, p. 134; *Spaceflight*, July-Aug. 1977, p. 276

22. Skuridin, G.A., *op. cit.*, pp. 418-420; Blagonravov, A.A., et al., *op. cit.*, pp. 489, 599; Turnill, R., *The Observer's Book of Unmanned Spaceflight*, p. 136; Gatland, K., *op. cit.*, p. 132; *Astronautics and Aeronautics*, May 1963, p.134; *Cosmic Research*, Sep.-Oct. 1967, p. 565

23. Ponomarev, A.N., *The Years of the Space Era*, p. 194; Gurney, G. and C., *Cosmonauts in Orbit*, p. 147; Wukelic, G.E., *op. cit.*, p. 148; *Soviet Space Programs, 1966-1970*, p. 167; *Astronautics and Aeronautics*, May 1963, p. 134

24. *Astronautics and Aeronautics*, June 1964, p. 107

25. Daniloff, N., *Kremlin and the Cosmos*, p. 202; *TRW Space Log, 1975*, p. 51; *Soviet Space Programs, 1971-1975*, p. 94

26. *Spaceflight*, July-Aug. 1977, p. 276

27. *Spaceflight*, Nov. 1969, p. 378; *Soviet Space Programs, 1971-1975*, p. 560

28. Turnill, R., *op. cit.*, p. 136; Skuridin, G.A., *op. cit.*, pp.420-421; Blagonravov, A.A., et al., *op. cit.*, pp. 599-600; *Spaceflight*, Nov. 1969, pp. 378-384; Gatland, K., *op. cit.*, p. 133; *Spaceflight*, July-Aug. 1976, p. 274; Sobel, L.A., ed., *Space: From Sputnik to Gemini*, pp. 284-285

29. *Soviet Space Programs, 1971-1975*, p. 460; Blagonravov, A.A., et al., *op. cit.*, p. 600; Skuridin, G.A., *op. cit.*, pp. 422-424

30. *Soviet Space Programs, 1971-1975*, pp. 94, 561

31. Gatland, K., *op. cit.*, pp. 133-134; Skuridin, G.A., *op. cit.*, p. 600; *Spaceflight*, Feb. 1966, p. 61

32. UPI, Moscow, 10 Oct. 1965

33. *Spaceflight*, Mar. 1969, p. 110; *Soviet Space Programs, 1971-1975*, p. 562; Skuridin, G.A., *op.cit.*, pp. 426-428; Blagonravov, A.A., et al., *op. cit.*, pp. 600-601; Gatland, K., *op. cit.*, p. 134

34. Blagonravov, A.A., et al., *op. cit.*, p. 603; *Soviet Space Programs, 1971-1975*, p. 562. (For further analysis of Luna 4-9 flight trajectories see Blagonarov, A.A., et al., *op. cit.*, pp. 288-299; *USSR Explores the Moon*, pp. 11-13; *First Panoramic Views of the Lunar Surface*, pp. 18-20; *Spaceflight*, July-Aug. 1976, p. 276; *Interavia*, Mar. 1966, pp. 283-284.)

35. Wilding-White, T.M., *Jane's Pocket Book of Space Exploration*, p. 41; *USSR Explores the Moon*, pp. 8-13; *Spaceflight*, Apr. 1966, pp. 131-135 and Mar. 1969, pp. 110-111 and Nov. 1969, p. 379; Wukelic, G.E., *op. cit.*, p. 64; Blagonravov, A.A., et al., *op. cit.*, p. 601

36. *First Panoramic Views of the Lunar Surface*, pp. 7-8; Wukelic, G.E., *op. cit.*, p. 280; *USSR Explores the Moon*, p. 8; Blagonravov, A.A., et al., *op. cit.*, pp. 177, 180

37. *First Panoramic Views of the Lunar Surface*, p. 8; *USSR Explores the Moon*, pp. 8-9

38. Petrovich, G.V., *op. cit.*, p. 233; *Spaceflight*, Nov. 1969, p. 381; *First Panoramic Views of the Lunar Surface*, p. 4; *USSR Explores the Moon*, p. 6

39. *First Panoramic Views of the Lunar Surface*, pp. 4-7; *USSR Explores the Moon*, pp. 6-7

40. Lunar 9 is considered to have made a "soft" landing on the moon as defined by its ability to survive the landing. *Nature*, 9 Apr. 1966, pp. 150-151; Gatland, K., *op. cit.*, p. 134

41. Blagonravov, A.A., et al., *op. cit.*, pp. 601-604; Skuridin, G.A., *op. cit.*, pp. 438-450; Petrovich, G.V., *op. cit.*, pp. 233-234; Wilding-White, T.M., *op. cit.*, p. 41; *Spaceflight*, July-Aug. 1976, pp. 276-277.
(One of the few Soviet released pictures illustrating the landing probe can be found in *USSR Explores the Moon*, p. 34.)

42. Aleksayev, V. and Lebedev, L., *For Lunar Rock*, p. 54; Skuridin, G.A., *op. cit.*, p. 463; *USSR Explores the Moon*, pp. 9-10; *First Panoramic Views of the Lunar Surface*, pp. 10-15

43. Shelton, W., *op. cit.*, p. 226; *Astronautics and Aeronautics*, May 1966, pp. 40-50; Gatland, K., *op. cit.*, pp. 134-135.
(See *First Panoramic Views of the Lunar Surface* for more details associated with the flight and analysis of the photographs.)

44. *TRW Space Log, 1975*, p. 56; *Science News*, 19 Aug. 1967; Daniloff, N., *op. cit.*, p. 202; Shelton, W., *op. cit.*, p. 208; *Soviet Space Programs, 1971-1975*, pp. 96, 563

45. Skuridin, G.A., *op. cit.*, p. 489; *Soviet Space Programs, 1971-1975*, p. 563

46. Woods believes this to be rounded off from an actual 1565 kg (*Spaceflight*, Mar. 1969, pp. 110-111)

47. *Jane's All the World'a Aircraft, 1966-1967*, p. 473; Blagonravov, A.A., et al., *op. cit.*, p. 602; (See also *Spaceflight* July-Aug. 1976, p. 278.)

48. Skuridin, G.A., *op. cit.*, p. 492

49. Wukelic, G.E., *op. cit.*, pp. 66, 83, 87, 169, 280, 287; *USSR Explores the Moon*, pp. 17-22; Petrov, G.E., *Conquest of Outer Space in the USSR, 1967-1970*, pp. 216-218; Blagonravov, A.A., et al., *op. cit.*, pp. 255-257, 300-34; Skuridin, G.A., *op. cit.*, pp. 491-495; *Interavia*, May 1966, p. 665

50. *USSR Explores the Moon*, pp. 18-22; Blagonarov, A.A., et al., *op. cit.*, p. 303

51. Petrov, G.I., *op. cit.*, p. 218; Skuridin, G.A., *op. cit.*, pp. 489-491; Blagonravov, A.A., et al., *op. cit.*, pp. 255, 605; *Spaceflight*, July 1966, pp. 244-246.
It is unclear whether the side instrument compartments were jettisoned prior to the retro burn as was done with Luna 9.

52. Petrov, G.I., *op. cit.*, p. 216; Wukelic, G.E., *op. cit.*, p. 169; Skuridin, G.A., *op. cit.*, pp. 496-507

53. *Spaceflight*, Jan. 1967, p. 18; *Soviet Space Programs, 1971-1975*, p. 564; Petrov, G.I., *op. cit.*, p. 218; Skuridin, G.A., *op. cit.*, pp. 509-512

54. *New York Times*, 31 Aug. 1966, p. 7; Wukelic, G.E., *op. cit.*, p. 66; Blagonravov, A.A., et al., *op. cit.*, pp. 605-606; *Spaceflight*, Jan. 1967, p. 18; Skuridin, G.A., *op. cit.*, pp. 509-513; *USSR Explores the Moon*, p. 22

55. Shelton, W., *op. cit.*, p. 232; Turnill, R., *op. cit.*, p. 138; *Spaceflight*, Oct. 1968, p. 361 and July-Aug. 1976, pp. 278-279; *Soviet Space Programs, 1971-1975*, p. 97

56. Skuridin, G.A., *op. cit.*,pp. 513-514

57. There is a report of a possible attempt on 23 Sept. 1966, see *Spaceflight*, July-Aug. 1977, p. 276; *Pokoryenye Kosmosa*, p. 75; Petrovich, G.V., *op. cit.*, p. 236; *Soviet Space Programs, 1971-1975*, p. 564; *Spaceflight*, Mar. 1969, pp. 110-11

58. *Spaceflight*, Mar. 1967, p. 100 and Oct. 1968, p. 361 and Nov. 1969, p. 380; Blagonravov, A.A., et al., *op. cit.*, p. 606; Skuridin, G.A., *op. cit.*, pp. 519-524

59. Petrovich, G.V., *op. cit.*, p. 237; Shelton, W., *op. cit.*, p. 232; Blagonravov, A.A., et al., *op. cit.*, p. 606; Skuridin, G.A., *op. cit.*, pp. 524, 527; *Soviet Space Programs, 1971-1975*, pp. 97-98. (For details of the photogrpahic system see description of Zond 3 in *Atlas of Reverse Side of the Moon, Part 2*.)

60. Turnill, R., *op. cit.*, p. 138; Skuridin, G.A., *op. cit.*, pp. 524-525; Blagonravov, A.A., et al., *op. cit.*, pp. 606-607; *Soviet Space Programs, 1971-1975*, p. 99

61. Skuridin, G.A., *op. cit.*, pp. 526-527

62. Blagonravov, A.A., et al., *op. cit.*, pp. 490, 608; Skuridin, G.A., *op. cit.*, p. 532

63. Yefremov, Yu.I., *The Panoramic Views of the Lunar Surface, Volume 2*, pp. 5-6; *Spaceflight*, Nov. 1969, p. 380

64. Yefremov, Yu.I., *op. cit.*, pp. 8-9; Wukelic, G.E., *op. cit.*, p.314

65. Yefremov, Yu.I., *op. cit.*, pp. 9-12, 20-21, 26-27; Wukelic, G.E., *op. cit.*, p. 313

66. Yefremov, Yu.I., *op. cit.*, pp. 12-15, 27-29; Skuridin, G.A., *op. cit.*, pp. 538-541; Blagonravov, A.A., et al., *op. cit.*, p. 177

67. Ponomarev, A.N., *op. cit.*, p. 194; Alekseyev, A. and Lebedev, L., *op. cit.*, p. 54; Skuridin, G.A., *op. cit.*, p. 531; Blagonravov, A.A., et al., *op. cit.*, pp. 608-609

68. Petrovich, G.V., *op. cit.*, p. 238; *Spaceflight,* Aug. 1968, pp. 268, 272

69. *Astronautics and Aeronautics, 1967,* (NASA), p. 206; *Astronautics and Aeronautics, 1968,* (NASA), p. 83

70. *Spaceflight,* Aug. 1968, p. 273 and July-Aug. 1976, p. 279; *Astronautics and Aeronautics, 1968,* (NASA), p. 83; Petrov, G.I., *op. cit.*, p. 219; *Soviet Space Programs, 1971-1975,* p. 99

71. For more background see *Soviet Space Programs, 1971-1975.*

72. *Astronautics and Aeronautics, 1969,* (NASA), p. 195; *Time,* 25 July 1969, p. 17; *Spaceflight,* July-Aug. 1976, p. 274 and July-Aug. 1977, p. 277

73. Petrov, G.I., *op. cit.*, pp. 256-257; *Soviet Space Programs, 1971-1975,* pp.145, 575. There is some dispute over the exact initial lunar orbit parameters. They may have been 870 km by 240 km.

74. Gatland, K., *op. cit.*, p. 146; Petrov, G.I., *op. cit.*, pp. 257-259; *Spaceflight,* Nov. 1969, p. 380

75. *TRW Space Log, 1975,* p. 73; *Spaceflight,* Feb. 1976, p. 76; *Soviet Space Programs, 1971-1975,* p. 575; *Mathematics Teaching,* No. 59, Summer 1972, p. 48

76. *Aviation Week and Space Technology,* 28 July 1969 and 17 Nov. 1969; *Astronautics and Aeronautics, 1969* (NASA), p. 195; *TRW Space Log, 1974,* p. 16; Sheldon, C.S., *United States and Soviet Progress in Space: Summary Data Through 1975 and a Look Forward,* p. 55; *Soviet Space Programs, 1966-1970,* pp. 363, 367, 368, 374; *Soviet Space Programs, 1971-1975,* p. 146

77. Lewis, R.S., *From Vineland to Mars,* p. 223; *TRW Space Log, 1975,* pp. 74-75; *Aviation Week and Space Technology,* 17 Nov. 1969 and and 10 Apr. 1972; *Spaceflight,* Jan. 1977, p. 40 and July-Aug. 1977, pp. 276-277 and Nov. 1977, pp. 414-415; *Soviet Space Programs, 1971-1975,* pp. 146, 576-577

78. *Newsweek,* 27 Apr. 1970, p. 112; *Spaceflight,* July-Aug. 1977, p. 277

79. *Soviet Space Programs, 1966-1970,* pp. 363-368

80. Alekesyev, V. and Lebedev, L., *op. cit.*, pp. 10, 21, 24; Petrov, G.I., *op. cit.*, pp. 267-268, 283; *Soviet Space Programs, 1971-1975,* p. 580

81. *Astronautics and Aeronautics, 1970* (NASA), pp. 299-300; Alekesyev, V. and Lebedev, L., *op. cit.*, pp. 26-33

82. *Spaceflight,* Feb. 1976, p. 76 and July-Aug. 1976, p. 280; Lewis, R.S., *op. cit.,* p. 223; *Soviet Space Programs, 1971-1975,* p. 148

83. Alekesyev, V. and Lebedev, L., *op. cit.,* pp. 13-14, 24-26; Petrov, G.I., *op. cit.,*pp. 277-279; *Spaceflight,* July-Aug. 1976, p. 280

84. Alekesyev, V. and Lebedev, L., *op. cit.,* p. 24; Petrov, G.I., *op. cit.,* p. 284; Wilding-White, T.M., *op. cit.,* p. 43

85. Petrov, G.I., *op. cit.,* p. 286; Alekesyev, V. and Lebedev, L., *op. cit.,* p. 27; *Aviation Week and Space Technology* 12 Oct. 1970; Turnill, R., *op. cit.,* p. 140; *Interavia,* Nov. 1970, p. 1346

86. Petrov, G.I., *op. cit.,* pp. 280-281; Wilding-White, T.M., *op. cit.,* p. 43; Alekesyev, V. and Lebedev, L.; *op. cit.,* pp. 14-15

87. Lewis, R.S., *op. cit.,* p.223; Petrov, G.I., *op. cit.,* p. 287; Alekesyev, V. and Lebedev, L, *op. cit.,* pp. 28-33

88. See the following for a more detailed analysis: *Astronautics and Aeronautics, 1970* (NASA), pp. 316, 358-359, 369; Alekesyev, V. and Lebedev, L., *op. cit.,* pp. 36-40; Petrov, G.I., *op. cit.,* pp. 288, 297-303; *Investigations of the Upper Atmosphere and Outer Space Conducted in 1970 in USSR (14th COSPAR),* pp. 2-10; Smolders, P., *Soviets in Space,* p. 204

89. Vassiliev, M., *Sputnik into Space,* pp. 126-131; Zaehringer, A.J., *op. cit.,* pp. 146-147; Rebrov, M. and Khozin, G., *op. cit.,* pp. 103-105; *Aviation Week and Space Technology,* 18 Jan. 1969, pp. 28-29

90. *Space World,* Apr. 1971, pp. 25-27; Petrov, G.I., *op. cit.,* p. 305; *Soviet Space Programs, 1971-1975,* p. 582

91. Alekesyev, V. and Lebedev, L., *op. cit.,* p. 77

92. Alekesyev, V. and Lebedev, L., *op. cit.,* pp. 48-49; *Space World,* Apr. 1971, p. 36; Turnill, R., *op. cit.,* p. 143; *Aviation Week and Space Technology,* 30 Nov.1970; Sheldon, C.A., *op. cit.,* p.56; Ponomarev, A.N., *op. cit.,* p. 270

93. Ponomarev, A.N., *op. cit.,* pp. 271-272; *Space World,* Sept. 1971, pp. 15-17; Smolders, P., *op. cit.,* pp. 210-212; Alekesyev, V. and Lebedev, L., *op. cit.,* pp. 50, 52, 93

94. Alekesyev, V. and Lebedev, L., *op. cit.,* pp. 54-55; Turnill, R., *op. cit.,* p. 144

95. Alekesyev, V. and Lebedev, L., *op. cit.,* pp. 63, 103; Smolders, P., *op. cit.,* p. 206; Turnill, R., *op. cit.,* pp. 147-148

96. *Space World,* Sep. 1971, p. 17; Alekesyev, V. and Lebedev, L., *op. cit.,* pp. 63-64

97. Alekesyev, V. and Lebedev, L., *op. cit.*, pp. 64-68; *Aviation Week and Space Technology*, 7 Dec. 1970

98. Alekesyev, V. and Lebedev, L., *op. cit.*, pp. 59-60

99. Alekesyev, V. and Lebedev., *op. cit.*, p. 62; *Space World* Apr. 1971, p. 46

100. *Aviation Week and Space Technology* 30 Nov. 1970; Petrov, G.I., *op. cit.*, pp. 310-320; Alekesyev, V. and Lebedev, L., *op. cit.*, pp. 69-70, 86

101. *Astronautics and Aeronautics, 1970* (NASA), pp. 364-365; Alekesyev, V. and Lebedev, L., *op. cit.*, pp. 74-78; Petrov, G.I., *op. cit.*, pp. 305-308

102. Alekesyev, V. and Lebedev, L., *op. cit.*, pp. 79-83; Petrov, G.I., *op. cit.*, pp. 309-313

103. Alekesyev, V. and Lebedev, L., *op. cit.*, pp. 84-90; Petrov, G.I., *op. cit.*, pp. 314-322

104. *Astronautics and Aeronautics, 1971* (NASA), pp. 220, 272; Alekesyev, V. and Lebedev, L., *op. cit.*, pp. 90-98

105. Alekesyev, V. and Lebedev, L., *op. cit.*, pp. 100-101.
For further information in this area see: Alekesyev, V. and Lebedev, L., *op. cit.*, pp. 99-113; Smolders, P., *op. cit.*, pp. 205-214; *Soviet Space Programs, 1971-1975*, pp. 151-155; *Astronautics and Aeronautics, 1971* (NASA), pp. 306-307; *Investigations of the Upper Atmosphere and Outer Space Conducted in 1970 in USSR*, pp. 2-15; *Russian Report to COSPAR 15th Session*, pp. 1-4a.

106. *Russian Report to COSPAR 15th Session*, p. 6; *Soviet Space Programs, 1971-1975*, pp. 156, 586; *Astronautics and Aeronautics, 1971* (NASA), p. 246

107. *Russian Report to COSPAR 15th Session*, p. 6; Gatland, K., *op. cit., cit.*, p. 158; *Aviation Week and Space Technology*, 11 Oct. 1971; Turnill, R., *op. cit.*, p. 140; *Soviet Space Programs, 1971-1975*, p. 156

108. *Space Research Conducted in USSR in 1974*, p. 27

109. *Russian Report to COSPAR 15th Session*, p. 6; *Soviet Space Programs, 1971-1975*, p. 586; Gatland, K., *op. cit.*, p. 158.
(Possible earlier attempt of heavy lunar orbiter in Feb. 1970, see *Spaceflight*, July-Aug. 1977, 276-277.)

110. *TRW Space Log, 1974*, p. 16; *Russian Report to COSPAR 15th Session*, p. 6; *Space Research Performed in USSR in 1972*, p2; *Space Research Conducted in USSR in 1973*, p. 3; *Astronautics and Aeronautics, 1971* (NASA), p. 267; Turnill, R., *op. cit.*, p. 142; *Spaceflight*, July-Aug. 1976, p. 283; *Soviet Space Programs, 1971-1975*, pp. 156, 586

111. *Russian Report to COSPAR 15th Session,* pp. 6-7; *Space Research Research Performed in USSR in 1972,* pp. 2-3; *Space Research Conducted in USSR in 1973,* pp. 17-18; *Astronautics and Aeronautics, 1972,*(NASA), pp. 38, 46, 94; *Astronautics and Aeronautics, 1973,* (NASA), p. 283. *Soviet Space Programs, 1971-1975,* p. 157; Gatland, K., *op. cit.,* pp. 158-159

112. *Astronautics and Aeronautics, 1972* (NASA), p. 337

113. *Jane's All the World's Aircraft, 1972-1973; Soviet Space Programs, 1971-1975,* pp. 157, 588

114. Alekesyev, V. and Lebedev, L., *op. cit.,* pp. 115-116; *Jane's All the World's Aircraft, 1972-1973*

115. Alekesyev, V. and Lebedev, L., *op. cit.,* p. 116; Turnill, R., *op. cit.,* p. 142; *Soviet Space Programs, 1971-1975,* pp. 157-158

116. *Aviation Week and Space Technology,* 6 Mar. 1972, p. 15; *Astronautics and Aeronautics, 1972* (NASA), pp. 52-43; Alekesyev, V. and Lebedev, L., *op. cit.,* p. 116; *Soviet Space Programs, 1971-1975,* p. 158

117. *Aviation Week and Space Technology,* 6 Mar. 1972 and 30 Aug. 1976; *Spaceflight,* July-Aug. 1976, p. 280; Gatland, K., *op. cit.,* pp. 161-162; *Astronautics and Aeronautics, 1972* (NASA), p. 71

118. *Space Research Performed in USSR in 1972,* pp. 4-6, 8; *Space Research Conducted in USSR in 1973,* p. 18; *Space Research Conducted in USSR in 1974,* pp. 26-28; *Astronautics and Aeronautics, 1972* (NASA), p. 179; *Space World,* Oct. 1972, pp. 39-40

119. *Aviation Week and Space Technology,* 22 Jan. 1973; *Soviet Space Programs, 1971-1975,* pp. 159, 593

120. *Jane's All the World's Aircraft, 1973-1974,* p. 647; *Space Research Conducted in USSR in 1973,* pp. 2-3; *Aviation Week and Space Technology,* 22 Jan. 1973; *Astronautics and Aeronautics, 1973* (NASA), p. 8

121. *Aviation Week and Space Technology,*12 Feb. 1973; *Space Research Conducted in USSR in 1973,* p. 3; *Jane's All the World's Aircraft, 1973-1974,* p. 647; Sheldon, C.S., *op. cit.,*p. 57; Wilding-White, T.M., *op. cit.,* p. 45; *Soviet Space Programs, 1971-1975,* p. 161

122. *Jane's All the World's Aircraft, 1973-1974,* p. 647; *Space Research Conducted in USSR in 1973,* p. 3; *Aviation Week and Space Technology,* 12 Feb. 1973; *Soviet Space Programs, 1971-1975,* pp. 160-163

123. *Aviation Week and Space Technology,*22 Jan. 1973; Turnill, R., *op. cit.,* pp. 142, 150; *Soviet Space Programs, 1971-1975,* pp. 159-161

124. Turnill, R., *op. cit.*, pp. 148-150; *Soviet Space Programs, 1971-1975*, p. 160.
For further details see: *Space Research Conducted in USSR in 1973*, pp. 25-26; *Astronautics and Aeronautics, 1973* (NASA), pp. 24, 30, 110-111, 139, 173; *Spaceflight*, July-Aug. 1976, pp. 282-283.

125. *Space Research Conducted in USSR in 1974*, p. 22; *Soviet Space Programs, 1971-1975*, pp. 163, 600; *Conquest of Outer Space in the USSR, 1974*, pp. 373-374

126. *Space Research Conducted in USSR in 1974*, pp. 22-24; *Soviet Space Programs, 1971-1975*, p. 163; *Conquest of Outer Space in the USSR, 1974*, p. 375

127. *Jane's All the World's Aircraft, 1975-1976*, p. 673; *Space Research Conducted in USSR in 1974*, p.22; *Spaceflight*, July-Aug. 1976, p. p. 282; *Soviet Space Programs, 1971-1975*, p. 163; *Conquest of Outer Space in the USSR, 1974*, p. 376

128. *Spaceflight*, July-Aug. 1976, pp. 282-283; *Soviet Space Programs, 1971-1975*, pp. 163-164

129. *Jane's All the World's Aircraft, 1975-1976*, p. 673; *Space Research Conducted in USSR in 1974*, p. 24-25; *Soviet Space Programs, 1971-1975*, p. 602

130. *Space Research Conducted in USSR in 1974*, p. 25; *Jane's All the World's Aircraft, 1975-1976*, p. 673; *Conquest of Outer Space in the USSR, 1974*,pp. 378-383

131. *Space Research Conducted in USSR in 1974*, p. 25; *Conquest of Outer Space in the USSR, 1974*, pp. 378-383

132. *Spaceflight*, Nov. 1976, p. 381 and Jan. 1977, p. 16

133. *Aviation Week and Space Technology*, 23 Aug. 1976, 30 Aug. 1976, 11 Oct. 1976, and 2 July 1979; *Spaceflight*, Nov. 1976, p. 382 and Jan. 1977, pp. 16-17

134. *Science News*, 28 Aug. 1976, p. 134 and 18 & 25 Dec. 1976, pp. pp. 390-391; *Spaceflight*, Nov. 1976, p. 382 and Jan. 1977, p. 17; *Sky and Telescope*, Dec. 1976, p. 421; *Science News*,10 Dec. 1977, p. 390; *Spaceview*, Aug./Oct. 1976, pp. 46/166-49/169

135. *Flight International*, 27 May 1978, p. 1643; *Aviation Week and Space Technology*, 8 May 1978, p. 56; *Spaceflight*, Apr. 1978, p. 153, July 1978, p. 242, and Sep.-Oct. 1978, p. 347

136. *Space World*, Apr. 1971, p. 27 and Dec. 1970, p. 22; *Soviet Space Programs, 1966-1970*, pp. 387-388; *Soviet Space Programs, 1971-1975*, p. 525.

Section II
EXPLORATION OF THE MOON
ZOND SERIES

INTRODUCTION

There has been intense speculation concerning the question of whether the Soviet Union ever had a program to land men on the moon by 1970. To this day the question remains unanswered. However, the Soviets did conduct a series of flights with spacecraft which admittedly were capable of carrying cosmonauts on elliptical circumlunar flights. Although these unmanned flights appeared to be successful, no cosmonauts were ever allowed to go along for the ride. It is thought that the success of the US Apollo 8 lunar orbit mission brought a premature end to the Zond program.

The Zond (meaning "probe" in Russian) program originated in 1964 to disguise early Venera and Mars spacecraft attempts. Zond 1 was sent in the direction of Venus while Zond 2 and Zond 3 followed trajectories to the vicinity of Mars. It is likely that when the USSR decided to begin its unmanned testing of spacecraft capable of carrying human crews, they chose the Zond designator to avoid the attraction a new program would draw. This would also allow the eventual use of Zond with manned flights while leaving the Luna program specifically for unmanned spacecraft investigation of the moon itself.

The chart in Figure 1 lists those flights or attempted launchings considered to be related to the Zond manned lunar program. Although Zond 3 was of an earlier generation spacecraft and not directly related to the manned program, since it returned valuable data about the moon (particularly the good photographs of the far side of the moon), the flight is discussed here as well as in the Mars section of this book.

ZOND PROBE SUMMARY

Designator	Launch Date	Mission	Weight	Results
Zond 3	18 Jul 1965	Lunar far side photographs	950 kg	Returned detailed photographs of far side of moon during flyby; continued into heliocentric orbit
Kosmos 146	10 Mar 1967	?	5000 ?	Failed to leave earth orbit; uncertain of success of mission
Kosmos 154	8 Apr 1967	?	5000 ?	Failed to leave earth orbit; uncertain of success of mission
Unannounced	22 Nov 1967	?	?	Possible launch failure with spacecraft never achieving earth orbit
Zond 4	2 Mar 1968	Circumlunar type test?	5600 ?	No reports after launch; entered heliocentric orbit
Unannounced	22 Apr 1968	?	5600 ?	Possible launch failure with spacecraft never achieving earth orbit
Zond 5	14 Sep 1968	Circumlunar flight	5800 ?	First return of spacecraft after lunar flyby; ballistic reentry; first Soviet water recovery; returned photographs of earth
Zond 6	10 Nov 1968	Circumlunar flight	5800 ?	Circled moon; performed skip-jump reentry and landed in Soviet Union; returned photographs of earth and moon
Unannounced	8 Jan 1969	Circumlunar flight	5800 ?	Possible launch failure with spacecraft never achieving earth orbit
Zond 7	8 Aug 1969	Circumlunar flight	5800 ?	Repeat of Zond 6; first color photographs
Zond 8	20 Oct 1970	Circumlunar flight	5800 ?	Returned in ballistic trajectory over North Pole; second Soviet water recovery
Kosmos 382	2 Dec 1970	?	19,000 ?	Performed major earth orbital maneuvers; believed to be related to Soviet manned lunar program

Figure 1

Chapter 1

ZOND SERIES: FIRST-GENERATION SPACECRAFT

ZOND 3

The announced launch of Zond 3 came as somewhat of a surprise to Western observers (see also Mars section of this book). The two previous spacecraft in this series, Zond 1 and Zond 2, had been planetary probes sent to investigate Venus and Mars, respectively. However, on 18 July 1965 when Zond 3 was launched, neither Venus nor Mars was in the proper position for a spacecraft encounter. In retrospect Zond 3 had a dual mission: (1) investigate and photograph the far side of the moon and (2) test spacecraft systems for later Venus and Mars flights. Although Zond 3 was not a part of the Zond 4-8 man-oriented lunar program to which this chapter is mainly devoted, Zond 3 did furnish valuable data about the moon and will be detailed here.

Launched by the augmented Vostok rocket (A-2-e), Zond 3 entered a temporary earth orbit of 210 km by 164 km at 64.78°. The spacecraft was fired from this orbit into a heliocentric trajectory which would eventually take the probe close to the orbit of Mars. Since a rather large initial velocity is required to totally escape from the influence of the earth (11.2 km/sec), Zond 3 reached the orbit of the moon very quickly (approximately thirty-three hours).[1]

The Zond 3 spacecraft was virtually identical with the earlier Zond 1-2 and Venera 2-3 (Figure 2). These spacecraft in turn were only slightly improved versions of Mars 1 launched in November 1962. Outwardly, the differences can be seen in the slightly altered solar panels and hemispherical radiators. Zond 3 weighed about 950 kg and measured approximately 3.5 meters in length and approximately 4 meters across the solar panels. The 2-meter diameter parabolic antenna was designed for communications at near-planetary distances.[2]

The major experiment on board Zond 3 consisted of a photographic system by which detailed photographs of the moon's far side were to be obtained.

Figure 2 The Zond 3 spacecraft which photographed the far side of the moon in July 1965 before flying to near the orbit of Mars

Although quite similar in operation to the system used by Luna 3 in 1959, Zond 3's television equipment was much improved. Photographs were to be taken in the visible range while simultaneously obtaining lunar surface spectra in the 2500-3500 Å range and spectrophotometric data in the 1900-2700 Å range and in the 3-4 μ range. Included here is a diagram of the phototelevision device used by Zond 3 (Figure 3).[3]

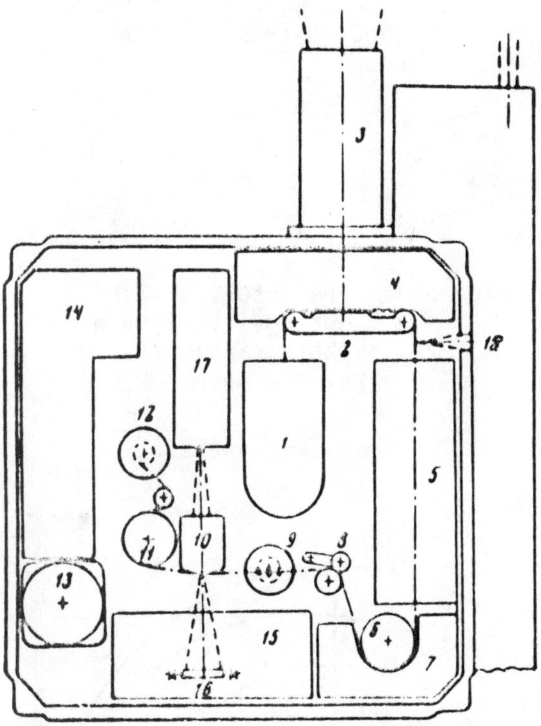

1. Contains 25.4 mm film
2. Film stand
3. Objective lens (focal length 106.4 mm)
4. Shutter
5. Developing device
6. Hot drying drum
7. Moisture absorber
8. Rollers
9. Film holder
10. Film window
11. Moving drum
12. Film holder
13. Motor which drives moving drum (11)
14. Electronic unit for photomultiplier
15. Optical mechanical scanning device
16. Scanning mirror
17. Photomultiplier
18. Ultraviolet spectrograph (for frames 8, 9, 10)

Figure 3 Schematic diagram of Zond 3 photographic system

The 6.5-kg phototelevision apparatus exposed a special 25.4-mm wide film through a series of conventional lenses and shutters. The film was then run through a developer, dried, and finally scanned by a television camera. Signals from the camera were then sent back to earth by one of two scan rates. First, all photographs were rapid-scanned (67 lines per frame) for immediate transmission to earth for quick analysis and determination of priorities for further scanning. In this mode one frame could be transmitted in 2 minutes 15 seconds. Later, all photographs were scanned in the normal mode (some were scanned several times). This time

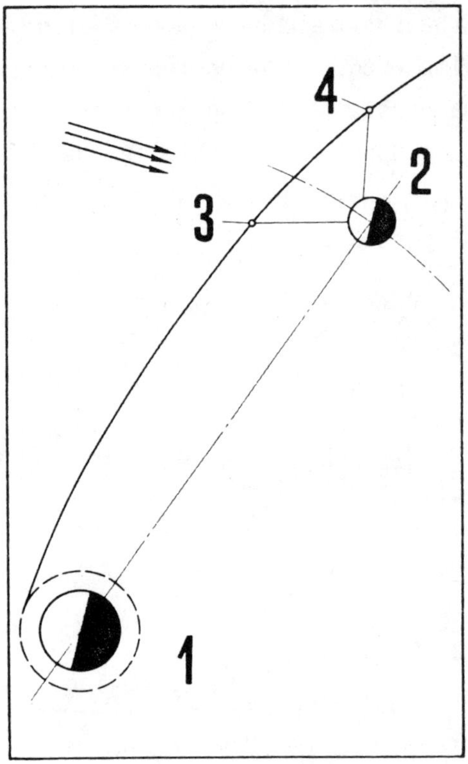

1. The earth
2. The moon
3. Start of the photographic session at 0424 on 20 July at a distance of 11,570 km from the lunar surface
4. Completion of the photographic session at 0532 at a distance of 9,960 km from the lunar surface

Figure 4a Trajectory of Zond 3 during its flyby and photographic session of the moon

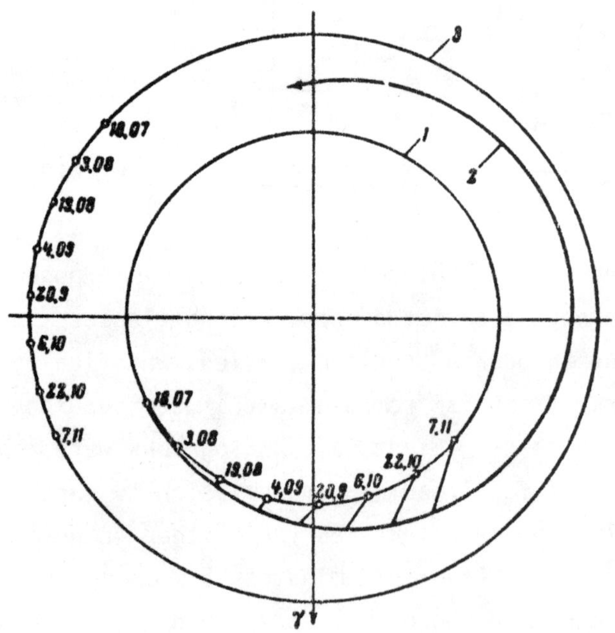

1. Orbit of the earth
2. Orbit of Zond 3
3. Orbit of Mars

Figure 4b Trajectory of Zond 3 after its encounter with the moon

each frame was broken up into 1100 lines and 860 elements per line for high quality and detail. Thirty-four minutes were required to transmit a single picture in normal scan.[4]

The photographic session began on 20 July 1965 at 0424 hr at a distance of 11,570 km above the moon. The angle of the sun with respect to the moon provided coverage of portions of both the near and far sides of the moon. Twenty-eight frames (three to the ultraviolet spectrograph) in all were exposed for periods of 1/100 and 1/30 sec with 2 minutes 14 seconds between exposures. During this time the spacecraft dipped to within 9,220 km of the lunar surface and then receded again to a distance of 9,960 km when the cameras stopped at 0532 hr Moscow time (Figure 4).[5]

The trajectory of Zond 3 was programmed such that the spacecraft would photograph the roughly 30% of the unseen side missed by Luna 3 (Figure 5). After review of the photographs it was determined that only 1.5 x 10^6 km^2 (approximately 5%) of the entire moon remained to be mapped by Soviet spacecraft. In addition, the flight was arranged such that the sun was illuminating the prime subject areas at a relatively low elevation. In this way the lengths of shadows cast by mountains and crater walls could be used to determine approximate heights.[6]

Figure 5 Photograph of the far side of the moon taken by Zond 3

The photographs were transmitted to earth at least twice: once on 29 July 1965 at a distance of 2.2×10^6 km and again much later at a distance of 31.5×10^6 km. Communications were maintained with Zond 3 until March 1966 at a distance of 150×10^6 km.[7]

Zond 3 carried a variety of other experiments in addition to the photographic equipment. Instruments on board the spacecraft monitored micrometeor flux, solar and cosmic rays, interplanetary as well as terrestrial magnetic fields, low-frequency radio emissions (in particular from the planet Jupiter) and the solar wind. Also plasma engines of the type used on Zond 2 were tested again.[8]

Chapter 2

SECOND-GENERATION MAN-ORIENTED SPACECRAFT

Between 1967 and 1970 the Soviet Union conducted extensive tests of an admitted manned lunar spacecraft system. With the peak of this activity centered in 1968, before the US Apollo 8 lunar orbit mission in December of that year, Soviet spokesmen acknowledged the capability of the system to carry a human crew on a highly elliptical earth orbit around the moon. No official acknowledgment of an eventual lunar orbit or lunar-landing capability for the Soyuz-derived spaceraft was made.

During the course of these test flights a variety of reentry and landing techniques were evaluated. First, a simple ballistic return trajectory with a touchdown in the Indian Ocean (the first ever Soviet water recovery) was performed. But, the resulting high g's associated with this technique rendered such a return undesirable for a cosmonaut crew. Later unmanned missions perfected a carefully controlled reentry approach of low g-loads and with recovery on Soviet territory.

Whatever the reasons for its cancellation, the magnitude of this man-oriented lunar program attests to the one-time determination of the USSR to fly men at least around the moon.

KOSMOS 146 and KOSMOS 154

That portion of the Zond program designed to develop spacecraft capable of carrying Soviet cosmonauts to the moon apparently began in March and April 1967. During this period Kosmos 146 and Kosmos 154 were launched one lunar month apart, also marking the first time the Proton launch vehicle had been reported used for something other than the Proton 1-3 satellites from which it derives its name.[9]

Kosmos 146 was launched on 10 March 1967 into an earth orbit of 310 km by 190 km at an inclination of 51.5°. A Soviet publication lists two objects in orbit with decay taking place the next day. However, Sheldon states that a Zond payload separated from the final stage on 11 March with decay of the final stage on 19 March and no data on the decay of the Zond payload. The Royal Aircraft Establishment (RAE) satellite tables indicate that the payload itself probably split into two pieces on 11 March exactly one day after launch.[10]

One lunar month later (8 April 1967) Kosmos 154 was launched into an orbit of 232 by 186 km at 51.6°. Again the USSR publication lists two objects in orbit while Sheldon believes the Zond payload remained attached to the final stage, and the RAE lists the payload and final stage separately. Since the Soviet definition of space "object" includes "the launcher, fairings, and other elements" a true determination of the flight is most difficult. All three sources, though, list the decay of Kosmos 154 just two days later on 10 April.[11]

Just what have the above two flights to do with the Zond program? It appears that these were the initial tests of the Zond spacecraft. Here it should be noted that the Zond spacecraft was virtually nothing more than a modified Soyuz spacecraft sans orbital module and that the first manned earth orbital mission of Soyuz had yet to appear.

One cannot be sure whether these two probes were designed to leave low earth orbit for lunar distances or whether they were lower-order test vehicles used merely to examine launch performances and possibly earth orbital reentry characteristics. In any event, all evidence points to failure of both craft to fulfill their programs whatever they were. Sheldon suggests that if these were Zond precursor failures, a possible setback was dealt to the Soviet lunar program.

A third attempt to launch a Zond spacecraft in 1967 is sometimes referred to by specialists in this field. An unsuccessful launch on 22 November 1967 is said to have resulted in the spacecraft never reaching earth orbit. A launch on this date would have been appropriate for a circumlunar flight.[12]

ZOND 4

The first acknowledged test of the man-oriented Zond spacecraft came on 2 March 1968. On this date, Zond 4 was placed in an earth orbit of 206 km by 190 km at 51.5°. What happened to Zond 4 from this point is open to question with almost as many opinions as there are authors on the subject. Both Soviet spokesmen and Western observers agree on Zond 4's general mission: the "perfecting of the new systems and assemblies on board" related to manned lunar flight. From here, though, a wide range of conclusions begins.[13]

No further reports on the flight of Zond 4 after the announcement of its launch have been found.[14] In the interest of completeness, that announcement is printed below as it appears in *Conquest of Outer Space in the USSR, 1967-1970,* edited by G.I. Petrov:[15]

> Zond-4 in Flight
>
> In accordance with the space research program, an automated probe Zond-4 was launched in the Soviet Union on March 2, 1968.
>
> The automated probe was introduced into the predetermined trajectory from an intermediate orbit as an artificial earth satellite. According to the measurement data, the automated probe Zond-4 is flying along a trajectory close to that calculated.
>
> The aim of the flight is to explore the further regions of near-earth space, as well as perfecting the new systems and assemblies on board.
>
> The coordination and computation center is processing the information received.
>
> Pravda, March 3, 1968.

What the announcement simply says is that Zond 4 was first put into earth orbit and then transferred to a "predetermined" trajectory. Two Soviet works on the history of the Soviet space programs list Zond 4 as entering a heliocentric orbit.[16] The launch took place at a time in which the moon was almost 180 degrees from a position needed if Zond 4 were to fly around it. However, as Sheldon suggests, this would have presented an ideal opportunity to test the return of a spacecraft from a lunar distance without encountering the perturbing effects of the moon itself.[17] Reentry into the atmosphere from a deep space flight is much more difficult than that from a simple low earth orbital mission, and hence new reentry systems must be developed and tested. If Zond 4 was intended to return to earth, then the mission must have failed since Soviet writers say it entered a solar orbit. Why Zond 4 would purposely be sent on a heliocentric path to test lunar systems is unclear.

In a book by Ley in 1969, Zond 4 is thought not to have left its earth parking orbit, but to have been destroyed when it naturally decayed and reentered the atmosphere.[18] The RAE's data suggest that Zond 4 may have burned up in the earth's atmosphere after a flight of 6.8 days, leading some investigators to feel that an unsuccessful recovery attempt was made.[19] The RAE's 6.8 days fits well with a typical elliptical circumlunar flight of about seven days. Woods believes Zond 4 was perturbed into a solar orbit from a high earth orbit, whereas Smolders believes it landed in the Indian Ocean on 9 March after a trip around an imaginary moon.[20]

The interest which Zond 4 has generated stems from two factors: (1) the scarcity of information and lack of Soviet enthusiasm for the flight and (2) the belief that the failure of Zond 4 coming when it did (early 1968) set back the Soviet lunar program and perhaps altered the course of the "space race". A comparison of the Zond 4 launch announcement with that issued for Zond 5 in September 1968 shows a verbatim similarity with the exception of two sentences added to the Zond 5 announcement before the last line. These sentences are:

> Steady radio communications are being maintained with the probe Zond-5. According to telemetric data, all the systems and assemblies on board and the scientific equipment are functioning normally.[21]

Those familiar with Soviet launch announcements readily recognize this phrase, which was also included in the subsequent Zond 6-8 press releases. It was, however, missing from the Zond 4 report. The Soviet Union has a reputation for secrecy and sometimes supplying misleading information, but very rarely has she been caught in a knowing lie. It is possible therefore based upon the above information and the past practices of the USSR in dealing with space flights that Zond 4 experienced a major communications or power problem early in flight. The Soviets therefore could be as much in the dark about exactly what happened to Zond 4 as is the rest of the world. Whether Zond 4 was to have attempted the first flight to fly to and return from a lunar distance (later accomplished by Zond 5) is thus unknown.

ZOND 5

Although the next admitted flight of the Zond spacecraft did not come until September 1968, there are reports of an attempted launch in April of that year. Around 22 April 1968 a Proton rocket is rumored to have failed in an attempt to place the Zond spacecraft in earth orbit.[22] Regardless, five months later the Soviets were ready with a good launch vehicle and lunar probe.

Zond 5 was rocketed into a temporary earth orbit of 219 km by 187 km at 51.5° at 0042 hr on 15 September 1968. Within 67 minutes of launch the last stage of the launch vehicle fired, accelerating Zond 5 to a velocity close to 11 km/sec and placing it on a trajectory to the moon. The probe was then separated from the carrier rocket.[23]

Few actual details have been released concerning the Zond spacecraft used on Zond 4-8 mission. It consisted largely of a Soyuz-type instrument compartment with solar cell panels and a Soyuz-type descent module (Figure 6). The docking collar in place of a Soyuz orbital module is believed to have been designed for docking with a proposed Soviet lunar lander. A parabolic antenna was also added to the descent module to permit high quality transmissions at lunar distances. In general its weight is thought to have been about 5800 kg (which also happens to be close to the maximum payload the Proton rocket can send to the moon). With a length of 5.3 m and a diameter of 2.3 m the spacecraft would indeed closely resemble a Soyuz spacecraft sans orbital compartment.[24]

At 0611 hr on 17 September at a distance of 325,000 km from earth Zond 5 made the first of two course corrections to be carried out on its flight. The spacecraft was now on a course which would take it close to and then circle behind the moon and then return to earth after a total flight of roughly seven days. Zond 5 made its closest approach to the moon of 1950 km on 18 September, and after reemerging from the other side flew a ballistic trajectory back to earth.[25]

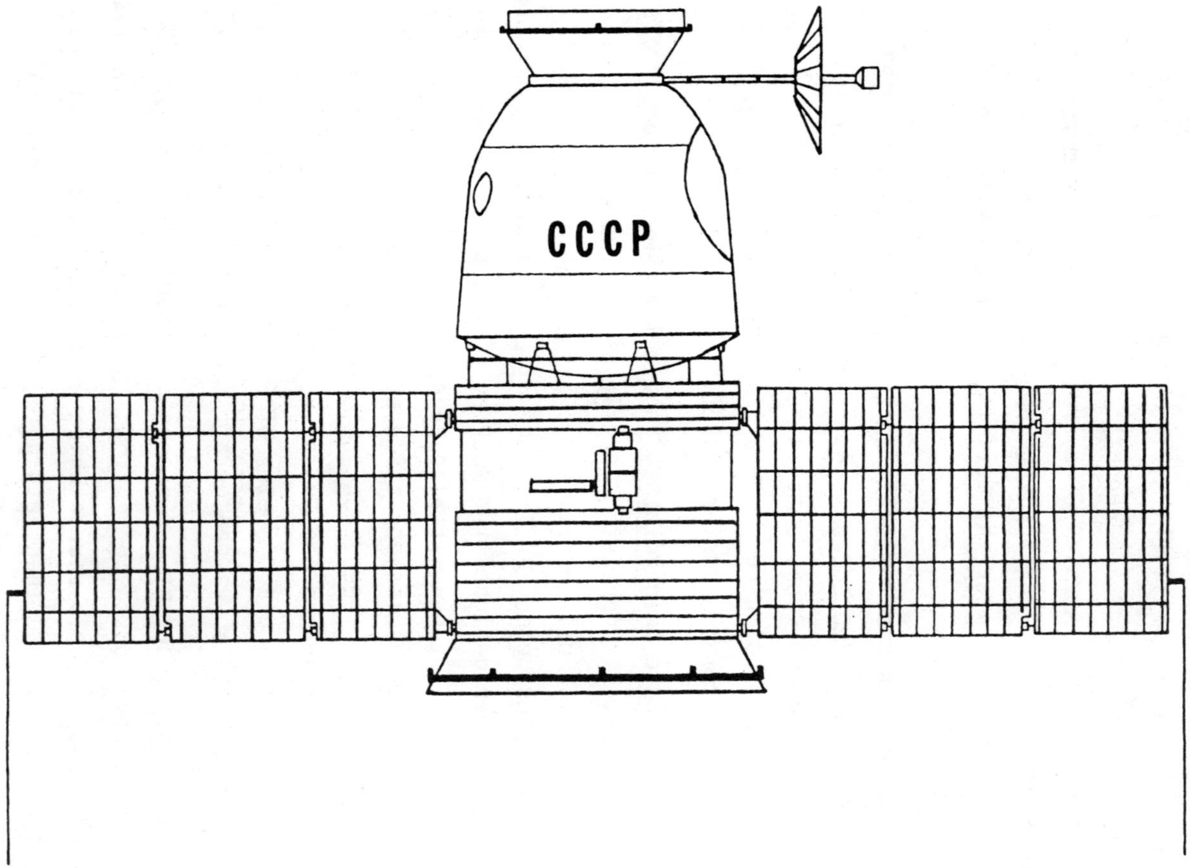

Figure 6   Sketch of the Zond spacecraft designed to carry men around the moon

Shortly before landing and at a distance of 143,000 km, Zond 5 performed a very small course-correction burn to ensure entry into the atmosphere at a carefully determined angle and landing at the designated location. Just before entering the atmosphere the descent module separated from the instrument module and rotated such that its heat shield was properly oriented with respect to the direction of travel. Reaching the atmosphere at an angle of 5-6 degrees to the horizontal at 1854 hr on 21 September with a velocity close to 11 km/sec, the deceleration forces quickly increased to 10-16 g's and the temperature of the heat shield reached 13,000°C. Hitting the atmosphere at an angle approximately one degree too steep would result in fatal loads on a human crew while entering the atmosphere at an angle approximately one degree too shallow would result in the spacecraft's skipping back out into space and possibly completing another loop around the moon.[26]

Fortunately, Zond 5 hit the 10-13 km wide entry corridor within the prescribed limits. At an altitude of 7 km with the velocity reduced to only 200 m/sec, the parachute system was deployed. By 1908 hr Zond 5 had safely performed the USSR's first water landing in the Indian Ocean at 32°38'S, 65°33'E. By the following day the Soviet ship Vasily Golovnin had located and recovered the descent capsule. On the eleventh anniversary of Sputnik 1, 4 October 1968, Zond 5 reached the port at Bombay, India, from which it was then flown to the Soviet Union.[27] Zond 5 thus returned to its starting point after having successfully completed the world's first circumlunar flight.[28]

Although just the successful return of Zond 5 represented a great technological feat, other interesting experiments were carried out during the flight. The most intriguing were reports that a tape recorder on board Zond 5 had been activated after its passage around the moon and sent back the voice of a "phantom" cosmonaut giving instrument readings. Apparently this was a test of the quality and reliability of the communications system designed for the manned flights which were to follow.[29] A wide variety of plant and animal life also tagged along for the ride, including an assortment of seeds (wheat, pine, and barley) and bacteria, a number of turtles, fruit flies, and mealworms which were sent to observe the long-term effects of solar and cosmic radiation. Several different types of instruments were used to record levels of radiation encountered during the flight.[30]

Finally, Zond 5 was equipped with a photographic system. However, this system did not require the elaborate processing and TV scanning of pictures as performed by Luna 3 and Zond 3. Since Zond 5 was returning to earth, the film could be processed here. For the first time, high-quality photographs from deep space were obtained. Whether Zond 5 attempted to photograph the moon during its flyby as did subsequent Zonds is unclear. A sharp, clear photograph of the earth taken at 1208 hr on 21 September from a distance of 90,000 km was released (Figure 7).[31]

Figure 7 Photograph of the earth taken by Zond 5 at a distance of 90,000 km during its return journey to earth

ZOND 6

Two lunar months after the Zond 5 flight, the Soviets were ready again, but this time a much more difficult maneuver was planned. Whereas Zond 5 made a simple ballistic reentry, Zond 6 was to attempt a controlled double entry return. A ballistic return from the moon had two major drawbacks for the USSR. The first was the excessive loads which would be placed on any crew (10-16 g's). Secondly, it was impossible to land on Soviet territory from a lunar ballistic trajectory. If the skip-jump approach was successful, not only would returning Soviet cosmonauts not be subjected to severe loads after being weightless for a long period of time, but also they would be able to guide their spacecraft aerodynamically, to land on their native soil.

On 10 November 1968 at 2211 hr Zond 6 entered an earth orbit of 210 km by 185 km at 51.4°, almost identical to the orbit flown by Zond 5. Again exactly like Zond 5, 67 minutes later Zond 6 was on its way to the moon. A mid-course correction was performed at 0841 hr on 12 November at a distance of 246,000 km from earth (Figure 8). With this new flight path Zond 6 whipped around the moon on 14 November at a minimum distance of 2.420 km. So far Zond 6 was an exact repeat of Zond 5.[32]

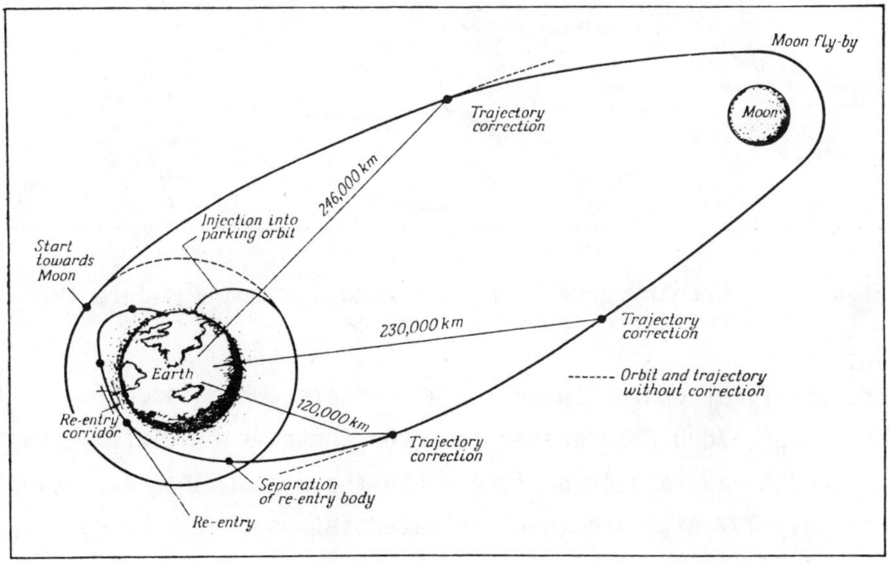

Figure 8 Trajectory of the Zond 6 spacecraft during its circumlunar flight and return to earth

On the return leg more emphasis was placed on the accuracy of Zond 6's trajectory. Two course corrections were made: the first at a distance of 236,000 km from earth on 16 November at 0940 hr and the second about eight hours before entry into the atmosphere at 0836 hr on 17 November while still 120,000 km from earth. Following these corrections Zond 6 was programmed to make its first plunge into the atmosphere through a 10-km wide corridor at an altitude of 45 km over the southern hemisphere, decelerating from 11 km/sec to 7.6 km/sec, then, by employing the aerodynamic features of the descent module, bounce back into outer space. During this brief reentry into space the spacecraft would travel several thousand kilometers before plunging back into the atmosphere for a second and final time (Figure 9).[33] The Zond 6 descent module did not have wing-like extensions to aid in reentry guidance as was sometimes reported at the time.[34]

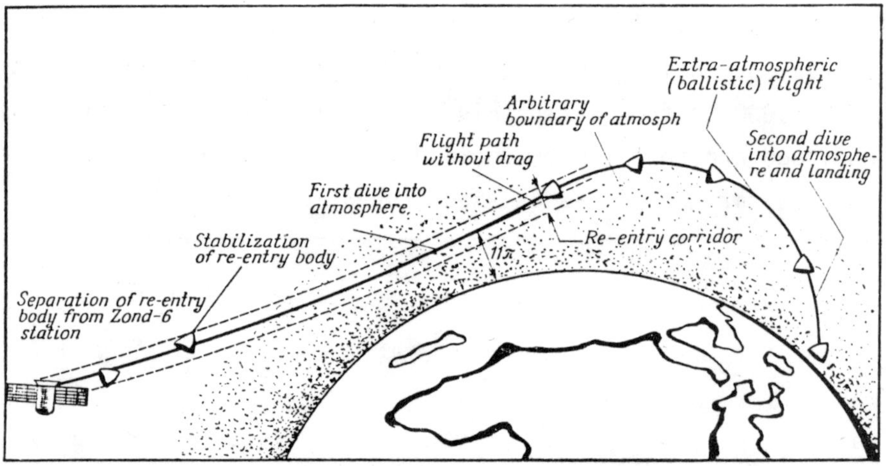

Figure 9 Reentry profile of the Zond 6 command module through the narrow reentry corridor

Prior to the first contact with the atmosphere, which occurred at 1658 hr on 17 November, Zond 6 separated from the instrument module. Exactly as planned Zond 6 made a searing foray into the atmosphere, experiencing loads of only 4-7 g's, and then retreated into space. The second plunge was very similar to that encountered by a spacecraft returning from earth orbit. Drogue chutes were released at an altitude of 7.5 km when the spacecraft velocity was about 200 m/sec. Landing took place a short time thereafter at an undisclosed location in the Soviet Union.[35] With the week-long mission completed, Zond 6 had proven that cosmonauts returning from the moon could safely land on Soviet territory--a requirement almost essential in Soviet eyes.

It was during the Zond 6 flight that the Soviets firmly announced that the Zond spacecraft was capable of carrying a human crew and was part of a manned lunar program.[36] The two impressive successes of Zond 5 and Zond 6 gave a burst of confidence to the Soviet space program and heightened public speculation of a Soviet manned circumlunar flight before the planned Apollo 8 moon mission.

Like Zond 5, Zond 6 carried a high-quality photographic system. This time, though, the unit was employed during the lunar flyby. Two photographic sessions were conducted with the first occurring at 0400 hr on 14 November as Zond 6 approached the moon from a distance of 11,000 km. These photographs pictured an almost fully illuminated lunar surface. As Zond 6 began its swing behind the moon at 0548 hr on 14 November at an altitude of 3,300 km the camera was again activated. However, this time a beautiful picture of the moon's limb with a half-earth in the background was taken (Figure 10).[37]

The photographs were taken by a camera with a focal length of 400 mm and isopanchromatic film with the dimensions 19 cm wide and 28.5 m long. Each frame was only 13 cm by 18 cm. Although the exact number of photographs taken was not released, Soviet writers did indicate that the number was much greater than the 25 pictures returned by Zond 3. Again much emphasis was placed on the greatly increased quality of pictures returned to and developed on earth over photographs which must be returned by facsimile transmission.[38]

An unspecified biological payload was carried on board Zond 6 as had been done on Zond 5. In addition, a photo-emulsion camera was employed to record the exact paths of cosmic rays, and a micrometeor detector was affixed to the surface of Zond 6. The latter system was put to good work as the returning spacecraft passed through the Leonid meteor shower.[39]

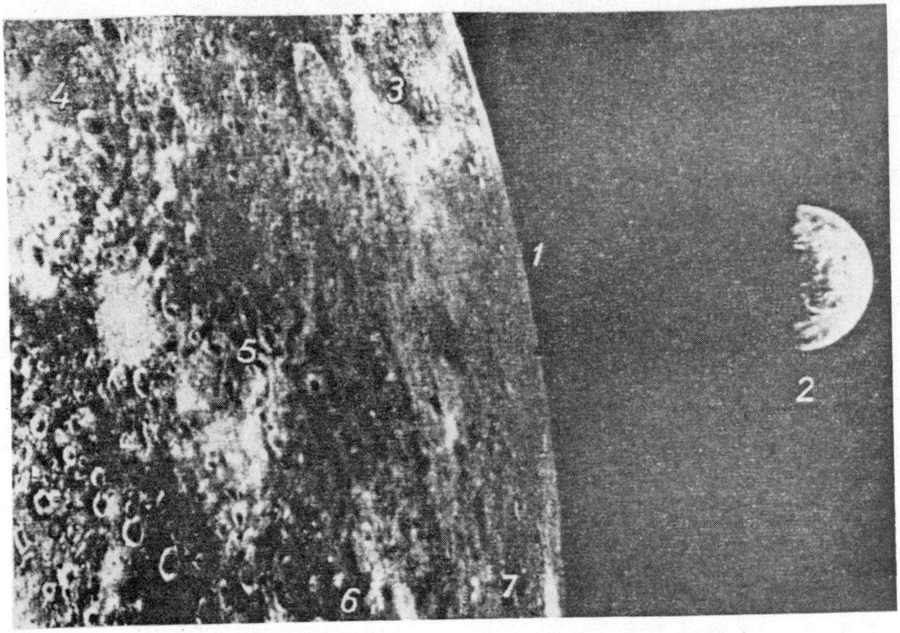

1. Edge of the moon
2. The earth
3. Sternberg
4. Lorentz
5. Langevin
6. Rynin
7. Buffon

Figure 10 Photograph of the earth and the moon's edge taken by Zond 6

ZOND 7

As already noted, the successes of Zond 5 and Zond 6 in September and Nov-November of 1968 led to increased speculation of manned Zond circumlunar flights. Many reports indicate a manned flight around the moon was scheduled for early January, 1969. In fact, one report suggests that an attempt may have been made to accelerate the flight schedule to December 1968 to beat the Apollo 8 mission. For whatever reasons, this latter event obviously did not occur. The January 1969 launch that was apparently attempted failed when the second stage exploded in mid-air. Evidently no Soviet cosmonauts were on board since a simple flyby of the moon would look second-rate compared with the ten lunar orbits performed by Apollo 8 only a few weeks before.[40]

The Zond program did not resume until two weeks after the historic landing of Apollo 11 on the moon. Up to this time, Soviet cosmonauts were still frequently speaking of a Soviet manned lunar landing by the end of 1969 or early 1970, although this would seem unduly optimistic given the absence of any flight tests of lunar spacecraft hardware other than the return capsule.[41] In any event, the decision was made to continue testing the Zond spacecraft, and on 8 August 1969 Zond 7 was launched.

After leaving an earth parking orbit of 191 km by 181 km at 51.5°, Zond 7 proceeded to repeat every aspect of the Zond 6 flight. Due to an apparent confidence in the photographic system, Zond 7 photographed the earth on the outward leg of its journey from 0532 hr to 0926 hr on 8 August at a distance of 70,000 km. In addition to black and white pictures like those taken by Zond 5 and Zond 6, Zond 7 was equipped to take the first Soviet deep-space color photographs. A later mid-course maneuver was performed on 9 August at a distance of 260,000 km.[42]

Zond 7 continued with a second photographic session with the moon as its subject from 0528 - 0538 hr on 11 August as the spacecraft rushed toward the moon from 10,000 km away. As Zond 7 began its swing around the moon about an hour and half later (0708 hr), it duplicated the feat of Zond 6 by capturing the earth and the edge of the moon in the same frame, but this time in color. After snapping the earth's picture Zond 7 made its closest approach to the moon (something less than 2000 km) and began the return journey home.[43]

Following a trajectory correction on the return flight, Zond 7 duplicated almost exactly the double-controlled entry into the earth's atmosphere and landed in the Soviet Union south of the town of Kustanai on 14 August. For the first time, mention was made of the use of retro-rockets similar to those employed by returning Soyuz spacecraft, approximately one meter above the ground to soften the impact. Equipment on board Zond 7 included:

1. System of motion control by using the spaceborne computer, ensuring the optimal control expression at all stages of the flight
2. Precision astro-orientation system
3. Equipment on board the probe for long-distance radio communications meant for receiving and transmitting information from the probe and for recording the parameters of its motion
4. Telemetric system for checking the functioning of the systems on board the probe

5. Devices for the protection of the spacecraft from radiation and for the control of the radiation dose in the landing vehicle.[44]

ZOND 8

The last flight of the Zond series came somewhat unexpectedly, just over a year after the return of Zond 7. Although Zond 8 differed from Zond 5-7 by using a third mode of reentry into the earth's atmosphere, the flight appears to have been an after-thought for the program. Certainly, even if the Soviets had had a manned lunar landing program, by this time it had been postponed indefinitely due to the failure to develop a large enough launch vehicle and due to the successes of the Apollo program. Thus, a flight at this late date was probably undertaken to take advantage of the existing hardware when the decision to terminate the Zond program was made, possibly in late 1969.

Zond 8 was launched on 20 October 1979 into an earth orbit of 223 km by 202 km at 51.5°. Before completing one revolution about the earth the final stage of the carrier rocket fired, placing Zond 8 on an elliptical trajectory designed to circle the moon. For the first time, Soviet press releases at the time of the launch specified exactly what the mission of Zond 8 was to be and when it would return to earth. On 21 October at a distance of 65,000 km, Zond 8 conducted its first photographic session of the earth. After a course correction the next day, Zond 8 swung around the moon at a minimum distance of 1120 km on 24 October and began the return leg of its journey.[45]

Whereas Zond 5 had made a ballistic reentry into the atmosphere after passing over the South Pole and Zond 6-7 had made a controlled aerodynamic reentry also after passing over that same region, Zond 8 was programmed to attempt a ballistic reentry after coming in over the North Pole. This enabled Soviet ground personnel to control the flight during the actual reentry. However, using this technique precluded the possibility of landing on Soviet territory and imposed large deceleration forces on the spacecraft and any prospective crew. The Zond 8 descent module plunged into the earth's atmosphere and made a searing ballistic reentry, landing approximately 730 km SE of Chagos Archipelego in the Indian Ocean at 1655 hr Moscow time on 27 October 1970. Although it was night in the Indian

Ocean when Zond 8 landed, this time the module was quickly spotted and recovered. The spacecraft was then transported by ship to Bombay where it was flown to the Soviet Union.[46]

Chapter 3

ZOND MANNED LUNAR PROGRAM

With the successful recovery of Zond 8, the USSR announced the conclusion of the Zond program.[47] The program had tested and proven the basic equipment needed to support a simple manned circumlunar flight. Where then were the manned missions for which this costly program was designed? Soviet officials had definitely announced the goal of this latest phase of Zond exploration was to carry men (or a single man) around the moon, and Belyayev himself appears to have been convinced that he would make such a circumlunar flight.[48]

The answer to this riddle is by no means clear. Certainly the remarkable voyage of Apollo 8 into lunar orbit and back had its effect on Zond. A simple flyby loop around the moon by one or more cosmonauts would appear second rate after the American feat. However, such a mission would still be expected if the Zond program was to eventually support manned lunar orbit and lunar landing missions.[49] Perhaps the Soviets elected to reserve judgment on the fate of the Zond program until they had seen the outcome of the initial Apollo lunar landing missions and while they continued to proceed with the third generation of unmanned lunar explorers. The subsequent successes of Apollo 11 and its followers were almost certainly responsible for the premature demise of the Soviet manned lunar program.

If the Zond program was originally intended to eventually land men on the moon and safely return them to earth, then additional components remained to be tested. Unless the composite spacecraft was to be mated in earth orbit following numerous launches employing the Proton rocket, a much larger launch vehicle was needed. Although several reports have surfaced regarding attempts to perfect such a rocket, this Saturn-V type vehicle has never been successfully flown (See Appendix I). In addition no Soviet lunar landing vehicle capable of carrying men is known to have been tested in earth orbit.[50]

The flights of Kosmos 379, 382, 398, and 434 in 1970 and 1971 are quite often linked to a Soviet lunar landing program.[51] All four of these craft performed extensive orbital maneuvers which can in principle be shown to approximate maneuvers associated with translunar insertion, entrance and exit from lunar orbit, and even simulation of lunar landing. In fact, the evidence so far assembled presents a very convincing case.

Before proceeding it should be recognized that any attempt to outline the Soviet manned lunar program, places the writer at a great disadvantage. The USSR still has not publicly admitted even the existence of a program comparable to the US Apollo program. In fact, there are quite a few observers of the Soviet space program who doubt that the Soviets actually had such a program. Parties on both sides of the issue can quote Soviet spokemen, point out technical capabilities, and list space achievements which support their respective views.

Therefore, in the absence of confirmation by the USSR (highly unlikely now ten years after Apollo 11) the question one must first decide for oneself is "Did the Soviets make a concerted effort beginning in the early 1960's to land men on the moon in the 1969-1970 time frame?". If after a review of the available evidence the answer is "no", then any further questions are meaningless. If, however, one is convinced that the Soviets did have such a plan, then logically the next step is to consider of what that plan may have consisted.[52]

Any attempt from this point on to deduce a Soviet lunar mission profile must be considered pure speculation. Although several spacecraft configurations proposed have been the result of careful analysis of the requirements of any lunar mission and of assumed Soviet capabilities, the reader is cautioned that every plan thus far proposed, regardless of apparent detail, has been conceived in the absence of verifying data.

One of the more serious investigations of Soviet manned lunar mission profiles has been put forth by David R. Woods. Woods has postulated one possible mission configuration which would have required three separate launches. The first by the yet-to-be-flown (successfully) super-rocket (G-1) would orbit the propulsion module required for the translunar insertion and lunar orbit burns. A Proton (D-1) launch vehicle would then place the lunar lander in earth orbit, followed by the launch of a Zond (Soyuz sans orbital module) spacecraft by the standard A-2 rocket with a crew of three on board. After earth-orbit assembly the combination would proceed along a flight plan similar to that followed by Apollo astronauts: that is by employing a lunar orbital rendezvous for the two cosmonauts returning from the lunar surface. With minor variations the mission could be accomplished with only two launches: one G-1 and one A-2 launcher.[53]

Phillip S. Clark, a noted Soviet space enthusiast from Britain, has also conducted an extensive review of Soviet lunar capabilities.[54] While he is in general agreement with the scenario proposed by Woods, he has suggested an alternate lunar mission which may have served as a prelude to a manned landing. Had the G-launch vehicle been made operational by the end of the 1960's, Clark sees a Zond/Salyut lunar orbit mission of value in order to allow the Soviets to gain lunar orbit experience. The flights of Apollo 8 and Apollo 10 did much the same for the United States. Under this scenario a three-man Zond would be placed in earth orbit by an A-2 rocket. Twenty-four hours later a Salyut type orbital station would be inserted on a rendezvous trajectory by a type G launcher. Following a quick docking maneuver and systems check out, the combination would be fired into a translunar trajectory (Figure 11). Upon reaching the moon Zond/Salyut might enter into a 110-km orbit and conduct a reconnaissance mission for a period of perhaps a month. The Zond spacecraft would undock and return to earth at the end of the planned program, leaving the Salyut station in lunar orbit.

In the event that a Zond/Salyut mission had been successful, the Soviets are hypothesized to switch emphasis to a Zond/lunar lander configuration similar to that proposed by Woods in his two launch vehicle concept. The final manned lunar landing attempt might follow a mission in which the lunar lander had been tested unmanned via remote control by cosmonauts in lunar orbit (Figures 12-16).

Although the preceding theories are of great value in evaluating possible Soviet lunar landing programs from the scant evidence available, mention should be made of the similarity of these schemes with that of the US Apollo program. Little serious thought in print has been given to the direct ascent and other various mission profiles by Western experts when

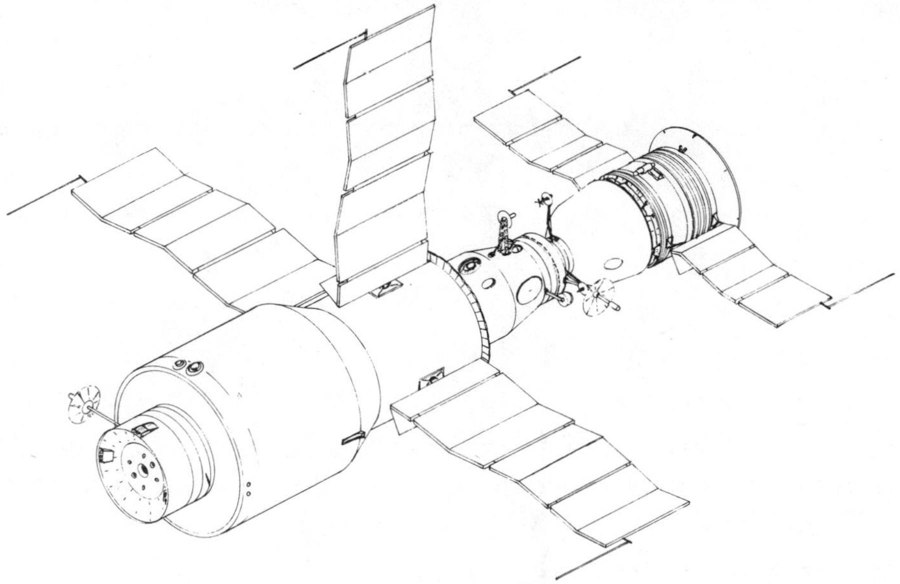

Figure 11 Speculative sketch of manned Zond spacecraft docked to a Salyut space station in lunar orbit (R.F. Gibbons)

postulating Soviet lunar plans. Although one is naturally inclined to think in terms of the familiar, it should be remembered that the lunar orbital rendezvous employed by Apollo was decided upon only after much heated debate. Also, the Soviet Union has time and again unveiled unique and astonishingly simple solutions to difficult problems.

What then can be said of future Soviet intentions regarding manned lunar exploration? Immediately before and after the Apollo 11 success, a Soviet landing by the mid-1970's was still predicted. The absence of this or any similar deep involvement with lunar exploration (only one unmanned Luna probe since 1974 although a far-side sample return mission is planned for the early 1980's) indicates that the Soviets may have temporarily suspended all manned lunar programs. The ongoing and extensive Soyuz/Salyut program along with the possible construction of a Soviet shuttle would probably leave a severely limited budget for a program of dubious worth at this stage of space exploration. However, continued Soviet interest in orbiting manned lunar laboratories and permanent lunar bases is apparent.

Scale 1/300

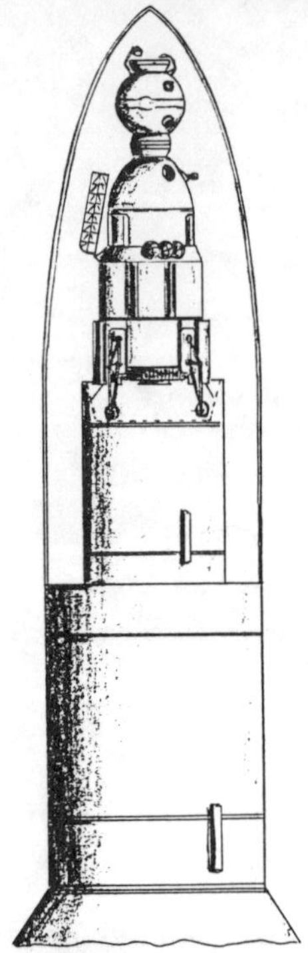

Figure 12 Proposed lunar landing payload perched atop the lunar transfer stage of the G-class launch vehicle (R.F. Gibbons)

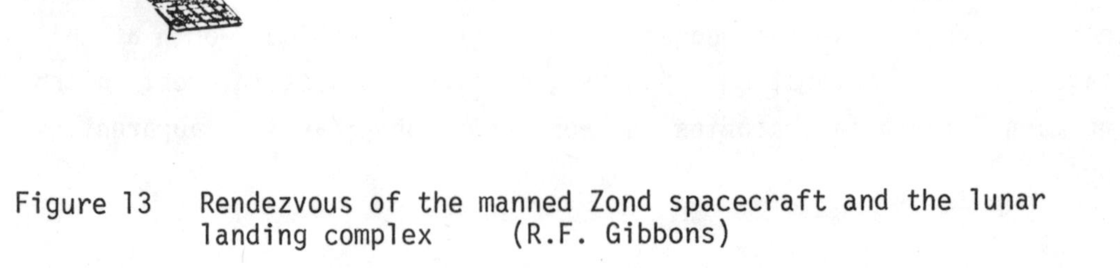

Figure 13 Rendezvous of the manned Zond spacecraft and the lunar landing complex (R.F. Gibbons)

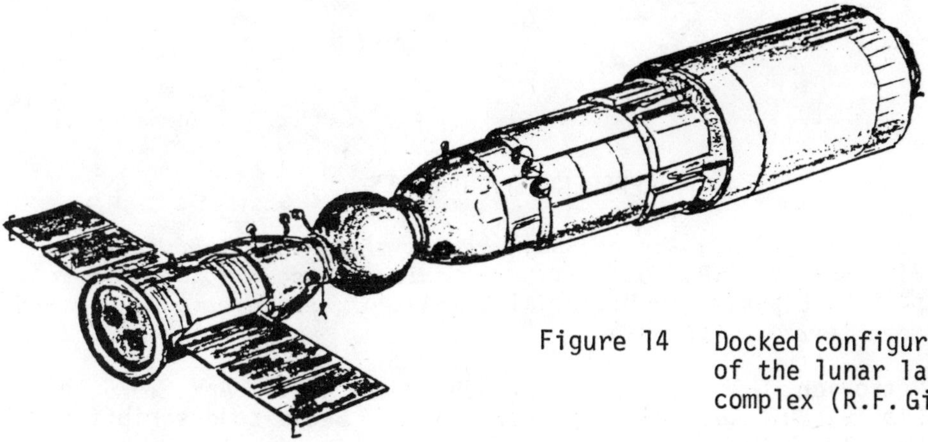

Figure 14 Docked configuration of the lunar landing complex (R.F. Gibbons)

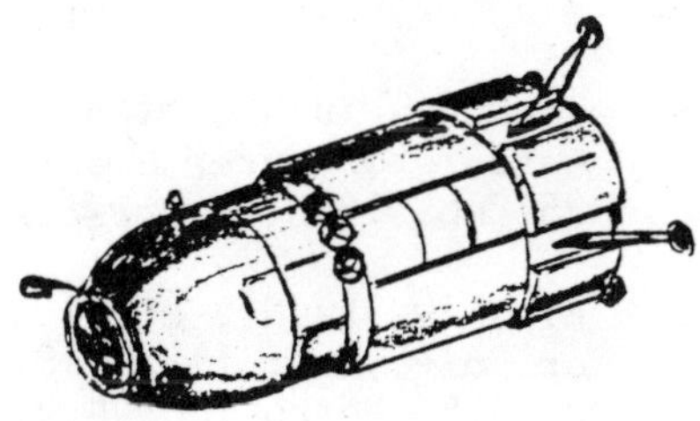

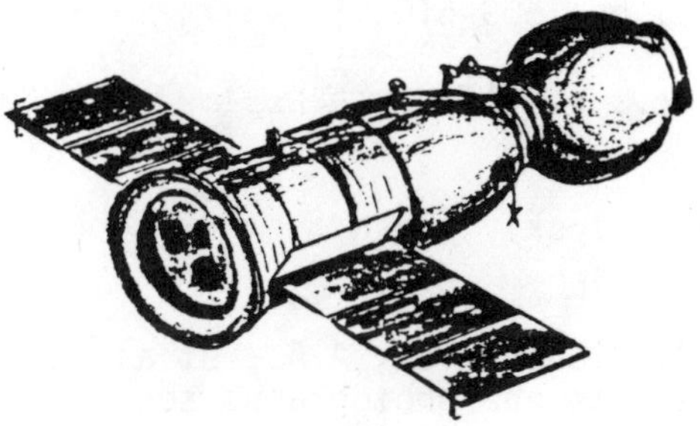

Figure 15 Undocking of the lunar lander from the Zond/Orbital Module (R.F. Gibbons)

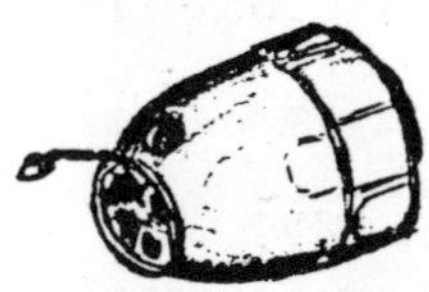

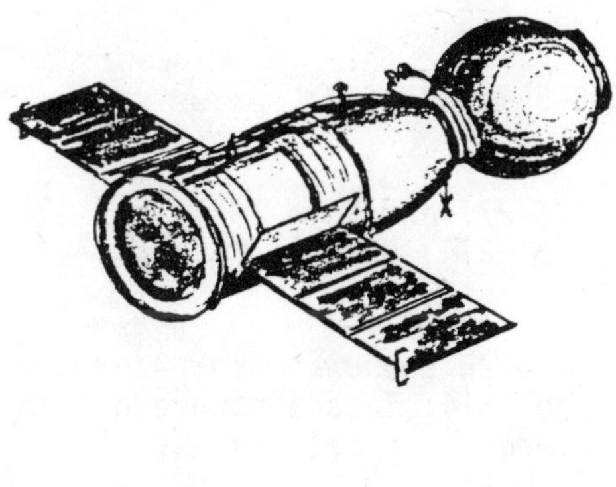

Figure 16 Rendezvous of the lunar lander ascent stage with the Zond/Orbital Module (R.F. Gibbons)

REFERENCES AND NOTES

1. Petrovich, G.V., *Soviet Encyclopedia of Space Flight*, p. 541; Blagonravov, A.A., et al.; *USSR Achievements in Space Research*, p. 615; Lipskiy, Yu.N., et al., *Atlas of the Reverse Side of the Moon, Part II*, pp. 3-6

2. Petrovich, G.V., *op. cit.*, p. 509; *Space World*, Apr. 1976, p. 21; Liskiy, Yu.N., et al., *op. cit.*, p. 8. See also descriptions of Mars 1 and Venera 2-3.

3. Blagonravov, A.A., *op. cit.*, pp. 282-285

4. Lipskiy, Yu.N., et al., *op. cit.*, pp. 11-16; Skuridin, G.A., *Mastery of Outer Space in the USSR, 1957-1967*, p. 1169; *Science*, 25 Feb. 1966; *Ten Years of Space Exploration*, p. 21

5. Lipskiy, Yu.N., et al., *op. cit.*, pp. 14-16; Skuridin, G.A., *op. cit.*, pp. 1183-1187; Blagonravov, A.A., et al., *op. cit.*, p. 615; Petrovich, G.V., *op. cit.*, pp. 509-510, 575-587

6. Blagonravov, A.A., et al., *op. cit.*, p. 615; *USSR Explores the Moon*, p. 5

7. Petrovich, G.V., *op. cit.*, p. 509; Lipskiy, Yu.N., et al., *op. cit.*, p. 11

8. Skuridin, G.A., *op. cit.*, p. 1179; Blagonravov, A.A., et al., *op. cit.*, p. 615. For further results and photographs see: Lipskiy, Yu.N., et al., *op. cit.;* Skuridin, G.A., *op. cit.*, pp. 1157-1195; Wukelic, G.E., *Handbook of Soviet Space-Science Research*, pp. 88, 279, 287

9. *Soviet Space Programs, 1971-1975*, pp. 57, 214, 565-566

10. Petrovich, G.V., *op. cit.*, pp. 556-557; *TRW Space Log, 1975*, p. 61; *Soviet Space Programs, 1971-1975*, p. 565

11. Petrovich, G.V., *op. cit.*, pp. 556-557, 572-573; *Soviet Space Programs, 1971-1975*, p. 566

12. *Spaceflight*, July-Aug. 1977, pp. 276-277 and Nov. 1977, pp. 414-415; *Soviet Space Programs, 1971-1975*, p. 214; Gatland, K., *Robot Explorers*, p. 140

13. Petrov, G.I., *Conquest of Outer Space in the USSR, 1967-1970*, p. 246; Petrovich, G.V., *op. cit.*, p. 514; Tass announcement in *New York Times*, 24 Nov. 1968; *Soviet Space Programs, 1971-1975*, p. 569

14. *New York Times,* 24 Nov. 1968. In addition no further reports have been found in all the publications and papers referenced in this book.

15. Petrov, G.I., *op. cit.,* p. 219

16. Petrovich, G.V., *op. cit.,* pp. 564-565; *Space World,* Apr. 1976, p. 21

17. *Soviet Space Programs, 1971-1975,* p. 215

18. Ley, W., *Events in Space,* p. 156

19. Private communication with Anthony Kenden, Sep. 1977.

20. *Spaceflight,* July-Aug. 1976, p. 285; Smolders, P., *Soviets in Space,* p. 222

21. Petrov, G.I., *op. cit.,* p. 220

22. *Spaceflight,* July-Aug 1977, pp. 276-277; *Soviet Space Programs, 1971-1975,* p. 214

23. Petrov, G.I., *op. cit.,* p. 225. See also *Spaceflight,* Feb. 1969, pp. 63-66 for a Soviet synopsis of the flight.

24. Petrov, G.I., *op. cit.,* p. 224; *Soviet Space Programs, 1971-1975,* pp. 215, 610

25. Petrov, G.I., *op. cit.,* p. 226

26. Petrov, G.I., *op. cit.,* pp. 221-228

27. Petrov, G.I., *op. cit.,* pp. 221-228; Petrovich, G.V., *op. cit.,* pp. 510-513

28. Contrary to popular belief the flights of Zond 5 and Zond 6 did not push U.S. authorities into a rush program to send Apollo 8 to the moon in December, 1968 to beat a possible Soviet manned flight. The tentative plans for Apollo 8 to make a circumlunar flight had been in the works since at least Aug. 1968. In fact final approval for the ten lunar orbit flight was made before the flight of Zond 5 was known. See *Aviation Week and Space Technology,* 23 Sep. and 30 Sep. 1968; *Spaceflight,* Dec. 1978, pp. 403-412.

29. Gatland, K., *op. cit.,* p. 141; *TRW Space Log, Winter 1968-1969,* p. 63; Turnill, R., *The Observer's Book of Unmanned Spaceflight,* p. 185; *Spaceflight,* Dec. 1968, p. 421

30. *TRW Space Log, Winter 1968-1969,* pp. 62-63; Petrov, G.I., *op. cit.,* p. 237; *Soviet Space Programs, 1971-1975,* p. 215

31. Petrov, G.I., *op. cit.,* pp. 230-232

32. Petrov, G.I., *op. cit.*, pp. 232-241

33. Petrov, G.I., *op. cit.*, pp. 238-244; Gatland, K., *op. cit.*, pp. 143-144; *Spaceflight*, Apr. 1969, pp. 135-136

34. *Jane's All the World's Aircraft, 1969-1970; Aviation Week and Space Technology*, 2 Dec. 1968; Petrov, G.I., *op. cit.*, p. 246

35. Petrov, G.I., *op. cit.*, pp. 242-244

36. Petrovich, G.V., *op. cit.*, p. 514; *New York Times*, 24 Nov. 1968; Petrov, G.I., *op. cit.*, p. 246

37. Petrov, G.I., *op. cit.*, pp. 252-256

38. Petrov, G.I., *op. cit.*, pp. 247-252

39 Petrov, G.I., *op. cit.*, p. 245

40. *Spaceflight*, July-Aug. 1976, p. 288 and July-Aug. 1977, p. 277 and Dec. 1978, pp. 403-412; Gatland, K., *op. cit.*, p. 144

41. *Soviet Space Programs, 1966-1970*, pp. 362, 368, 372, 374, 376-377, 390-392; *Soviet Space Programs, 1971-1975*, pp. 219-219; Daniloff, N., *The Kremlin and the Cosmos*, p. 168; James, P.N., *Soviet Conquest from Space*, p. 102; Sheldon, C.S., *United States and Soviet Progress in Space: Summary Data Through 1975 and a Forward Look*, pp. 80-81

42. Petrov, G.I., *op. cit.*, pp. 259-261; *Spaceflight*, Nov. 1969, p. 383, 395; *Soviet Space Programs, 1971-1975*, p. 576

43. Petrov, G.I., *op. cit.*, pp. 260-266

44. Petrov, G.I., *op. cit.*, pp. 260-261

45. Gatland, K., *op. cit.*, p. 150; Wilding-White, T.M., *Jane's Pocket Book of Space Exploration*, p. 47; *Soviet Space Programs, 1971-1975*, pp. 186-187, 506-508

46. Petrov, G.I., *op. cit.*, pp. 242, 305; *Soviet Space Programs, 1971-1975*, pp. 217-218

47. Petrov, G.I., *op. cit.*, p. 305; *Space World*, Apr. 1976, p. 21

48. Collins, Michael, *Carrying the Fire*, p. 282

49. There is some thought that the Zond spacecraft itself was capable of entering and leaving lunar orbit. *Spaceflight*, May 1975, p. 167

50. A suggestion that this has in fact occurred is found in *Spaceflight*, Nov. 1977, pp. 383-389.

51. *Spaceflight,* Apr. 1977, pp. 159-160 and Oct. 1973, pp. 398-400 and Aug.-Sep. 1975, p. 284 and Feb. 1974, p. 79; *Space View,* Jan.-Feb., pp 4/4; *Soviet Space Programs, 1971-1975,* pp. 186-187, 496-497, and 506-515

52. Some possible scenarios are given in the following: *Spaceflight,* Feb. 1974, p. 79 and May 1975, pp.183-191, 200; *Space World,* July 1977, pp. 26-31; *Soviet Space Programs, 1966-1970,* pp. 245-253, 390-393; *Soviet Space Programs, 1971-1975,* pp. 218-221, 502-515.

53. *Spaceflight,* July-Aug. 1976, pp. 283-284 and Oct. 1976, pp. 369-370 and Nov. 1977, pp. 383-388

54. Clark, Phillip S., *A Review of the Soviet Union's Manned Lunar Programme,* pp. 16-29

Section III
EXPLORATION OF VENUS

INTRODUCTION

The history of the Soviet exploration of the planet Venus is long indeed, spanning over eighteen years. Venera spacecraft were the first probes launched toward Venus and were the first (until 1978 the only) probes to return data from the atmosphere and the surface of Venus. The photographs returned by Venera 9 and Venera 10 are man's only glimpse of that eternally veiled planet. However, the road of exploration was not an easy one. Almost seven years and perhaps fourteen spacecraft passed from the launch of the first Venus probes in 1961 to the first return of scientific information from Venus by Venera 4. But determination and a relentless desire to unwrap the mysteries of our nearest planetary neighbor finally bore fruit in the successful flights of Veneras 4 - 12.

The chart of Figure 1 represents a summary of all known and suspected Soviet attempts to probe Venus. These data as with almost all the information in this book are either dependent upon disclosures by the Soviets or observations and conjectures by Western observers. However, with careful analysis of the given data these latter conjectures have, in many instances, been subsequently confirmed by representatives of the USSR.

As one reviews Figure 1 several important trends emerge. The first is the concerted effort of the Soviets in their exploration of Venus. Unlike the United States which launched only four spacecraft (Mariners 1, 2, 5, and 10) toward Venus between 1961 and 1975, the Soviets apparently launched or attempted to launch no less than twenty-three such craft. The only launch opportunity which was missed occurred in 1973 when the Venera series was evolving into its second generation of probes. In addition, all launchings occurred in at least pairs (twice three probes were launched and in 1964 as many as four Venus explorers may have been consumed). Each successive spacecraft drew on the successes and failures of its predecessors, resulting in more complex and versatile robot explorers. This culminated in the multi-purpose orbiter-lander combinations of Venera 9 and Venera 10.

VENUS PROBE SUMMARY

Designator	Date Launched	Mission	Total Weight	Atmospheric Probe Weight	Results
Sputnik 7	4 Feb 1961	Flyby	645 kg ?	----	Failed to leave earth orbit
Venera 1	12 Feb 1961	Flyby	643.5	----	Communications failed 27 Feb at a distance of 23 million km
Unannounced	25 Aug 1962	Flyby	890 ?	----	Failed to leave earth orbit
Unannounced	1 Sep 1962	Flyby	890 ?	----	Failed to leave earth orbit
Unannounced	12 Sep 1962	Flyby	890 ?	----	Failed to leave earth orbit
Kosmos 21	11 Nov 1963	Test ?	950 ?	----	Failed to leave earth orbit
Unannounced	27 Feb 1964	Flyby ?	950 ?	----	Apparent launch failure
Unannounced	4 Mar 1964	Flyby ?	950 ?	----	Apparent launch failure
Kosmos 27	27 Mar 1964	Flyby ?	950 ?	----	Failed to leave earth orbit
Zond 1	2 Apr 1964	Flyby ?	950	----	Communications failed about 14 May 1964
Venera 2	12 Nov 1965	Flyby	963	----	Communications failed Feb 1966 just before flyby
Venera 3	16 Nov 1965	Atmosphere Probe	960	?	Communications failed Feb 1966 just before entry into atmosphere
Kosmos 96	23 Nov 1965	Atmosphere Probe	960 ?	?	Failed to leave earth orbit
Venera 4	12 Jun 1967	Atmosphere Probe	1106	383	First return of data from within atmosphere on 18 Oct; crushed at altitude of 27 km
Kosmos 167	17 Jun 1967	Atmosphere Probe	1106 ?	383 ?	Failed to leave earth orbit
Venera 5	5 Jan 1969	Atmosphere Probe	1130	405	Entered atmosphere on 16 May; returned data to an altitude of 24-26 km
Venera 6	10 Jan 1969	Atmosphere Probe	1130	405	Entered atmosphere on 17 May; returned data to an altitude of 10-12 km
Venera 7	17 Aug 1970	Lander	1180	500	First successful landing on 15 Dec; returned data for 23 minutes from surface
Kosmos 359	22 Aug 1970	Lander ?	1180 ?	500 ?	Failed to leave earth orbit
Venera 8	27 Mar 1972	Lander	1184	495	Successful landing on 22 Jul; returned data for 50 minutes from surface
Kosmos 482	31 Mar 1972	Lander ?	1180 ?	500 ?	Failed to leave earth orbit
Venera 9	8 Jun 1965	Lander/ Orbiter	4936	1560	First Venerian orbiter 22 Oct; Lander returned first picture and data for 53 minutes from surface same day
Venera 10	14 Jun 1975	Lander/ Orbiter	5033	1560	Second orbiter 25 Oct; Lander returned picture and data for 65 minutes from surface same day
Venera 11	9 Sep 1978	Lander	4500 ?	1700 ?	Successful lander survived 95 minutes on surface on 25 December; no photos
Venera 12	14 Sep 1978	Lander	4500 ?	1700 ?	Successful lander survived a record 110 minutes on surface on 21 December; no photos

Figure 1

Chapter 1

FIRST-GENERATION SPACECRAFT

SPUTNIK 7, VENERA 1

The first attempt to send a probe to the vicinity of the planet Venus probably occurred on 4 February 1961 with the launching of Sputnik 7. This was the seventh Soviet satellite to orbit the earth since Sputnik 1, excluding Lunas 1 - 3 which followed direct paths to the moon in 1959. Although the Soviets have never admitted this satellite was a Venus probe, they have associated it with the development of "interplanetary ships".[1] In addition, its similarities with Venera 1 launched eight days later are convincing evidence of its probable purpose.

Sputnik 7 entered an orbit of 328 km by 224 km with an orbital inclination of 65°.[2] The orbital platform (probe plus Venus trajectory injection stage) weight has been reported at 6483 kg.[3] However, due to a malfunction of the orbital platform the actual Venus probe never left earth orbit.[4] The platform decayed and reentered the atmosphere on 26 February 1961.[5] While denying rumors that Sputnik 7 was an unsuccessful manned spacecraft, a Soviet official stated that the satellite carried no special scientific instruments and was concerned only with the problems involved in placing heavy loads into highly accurate orbits.[6]

On 12 February 1961 a second heavy Sputnik weighing 6475 kg was launched into an earth orbit of 319 km by 198 km with inclination of 65°.[7] Later that day the Venera 1 probe was launched from this orbital platform to become the first man-made vehicle to be aimed at the planet Venus (Figures 2 and 3). This was the first time that the Soviet technique of reaching earth orbit and then launching an interplanetary probe had been successfully accomplished after at least three failures.[8] The Venera 1 spacecraft which weighed 643.5 kg was separated from its satellite launcher after reaching a predetermined velocity.[9] Communications were maintained with the craft until 17 February 1961, but failed ten days later when

Figure 2 Photograph of the Venera 1 spacecraft launched in 1961

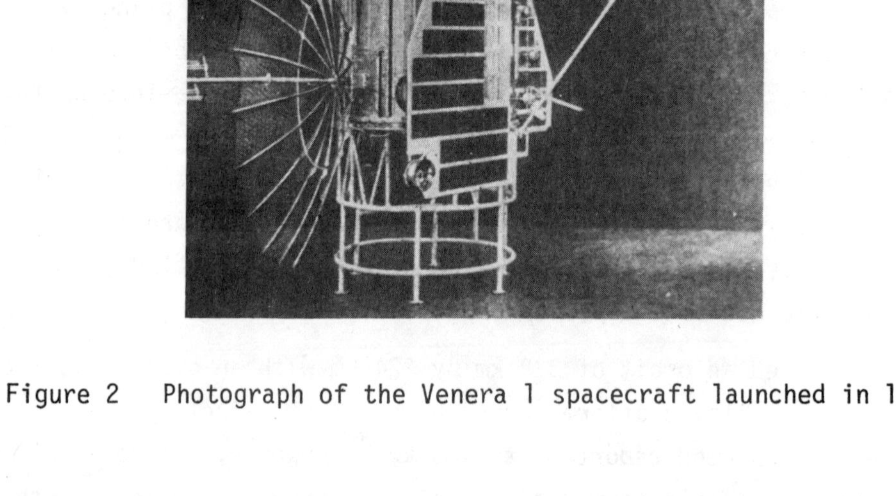

Figure 3 Diagram of the Venera 1 spacecraft

contact could not be reestablished at a distance of 23 million km from earth. The Soviet Union even solicited aid from the British Jodrell Bank radio telescope, but to no avail.[10] It is estimated that from its last known position Venera 1 passed Venus at a distance of approximately 100,000 km between 19 and 20 May 1961.[11]

Overall length and diameter of the interplanetary station excluding antennae and solar panels were 2035 mm and 1050 mm, respectively. For communications near earth an omnidirectional rod antenna 2.4 m in length was used, while two cross antennae were located on the solar panels to provide medium distance communications. On reaching the vicinity of Venus a high-gain parabolic mesh antenna with a diameter of approximately 2 m was to deploy automatically.[12]

To power the communications and other scientific equipment, chemical batteries were placed inside the center cylindrical container. For power and recharging of the batteries two panels of solar cells were constantly oriented toward the sun. These solar panels were approximately 1.6 m long and 1 m wide.[13]

The cylindrical section also contained part of the in-flight instrumentation and as a result a constant pressure of 900 mm Hg was maintained within this section. Temperature control was regulated by a system of internal and external shutters.[14] Among the instruments carried to study interplanetary space were a three component magnetometer to study the Venerian magnetic field and a variometer for measuring magnetic fields enroute. Also on board were solar-oriented charged-particle traps for measuring ion density of positive ions greater than 50 eV and electrons greater than 200 eV.[15]

The purpose of Venera 1 is generally assumed to have been that of flyby investigation, although some early reports purported that planetary impact was the goal.[16] To commemorate the first man-made probe to Venus a 70 mm diameter replica of the earth, made out of a titanium alloy and colored pale blue and golden-yellow was carried inside the spacecraft. Within the sphere lay a medal embossed with the emblem of the USSR.[17]

AUGUST-SEPTEMBER 1962 FLIGHTS

The launch windows by which spacecraft can be sent to Venus under minimum energy requirements occur approximately every 19 months. Therefore, the next opportunity following Venera 1 came in August and September of 1962. Whereas the USSR has never acknowledged any attempt to launch a Venera-type spacecraft during this period, the United States reported a launch took place on 25 August and left several pieces in orbit. Similar launches occurred on 1 September and 12 September, but again no probes were ever detected as having been separated from these platforms. The major portions of the satellites decayed in 3, 5, and 2 days, respectively.[18] These spacecraft probably resembled the Mars 1 probe launched less than two months later and to an extent the subsequent Zond 1-3 and Venera 2-3 probes

During this time the US and the USSR were accusing each other of secret launches and the procedure of reporting objects in orbit was revised. Soon after the newly begun series of Kosmos research satellites would become a catch-all for most military missions, spacecraft test flights, and unsuccessful lunar and planetary probes.

KOSMOS 21

On 11 November 1963 Kosmos 21 was rocketed into an earth orbit of 229 km by 195 km with an orbital inclination of 64.8°. The new satellite was almost certainly not a true planetary probe since no Mars or Venus launch window was open on this date. Kosmos 21's payload was unannounced and decayed within three days.[19] Charles Sheldon believes this to have been a test of spacecraft systems as a result of the several previous Venus and Mars probe failures.[20] A similar type test was performed by Zond 3 in July 1965 which, after photographing the far side of the moon, rendezvoused with an imaginary spot near the orbit of Mars. If Kosmos 21 was a test of the upcoming Venera explorers, it too was probably a failure since it did not leave earth orbit or make any further maneuvers.

KOSMOS 27 AND ZOND 1

When the February-April 1964 Venus launch window did arrive, the Soviets were ready. In the space of a month three Venera-type spacecraft are reported to have been launched, all of which ended in failure within hours.

On 26 February and 4 March the first two planetary probes appear to have suffered upper stage malfunctions and were immediately abandoned. A third try came on 27 March when Kosmos 27 entered an orbit remarkably close to that of Kosmos 21 (237 km by 191 km at a 64.8° inclination). However, no Venus probe was launched from this platform and Kosoms 27 decayed the next day.[21]

A fourth and final attempt six days later on 2 April was more successful with the earth orbit and subsequent injection into a Venus trajectory of Zond 1. Although few details of the spacecraft have been released, it has been reported that the probe was similar to that of Mars 1 and Veneras 2 and 3 with a weight in the neighborhood of 950 kg.[22] The chosen name of Zond (which in Russian means "probe") is viewed as an attempt by the Soviets to disguise the true purpose of the spacecraft if it encountered difficulties. As has already been seen, this program name was later taken by the related manned circumlunar missions.

During the course of the first month of the flight of Zond 1 about sixty communications were held with the spacecraft. Course corrections were performed at distances of 560,000 km and 13 million km. After seven weeks in space disaster struck when all communications with the spacecraft were lost shortly after 14 May 1964. Because of the inability to track Zond 1 further, estimates have predicted that the probe passed Venus at about 100,000 km or possibly even impacted on the planet around 19 July 1964.[23] Some scientific results of cosmic ray flux measurements have been reported from the early returned data.[24] Thus within three months four complicated and expensive spacecraft which had been programmed to investigate Venus had all failed.

VENERA 2, VENERA 3, AND KOSMOS 96

The fourth Venus launch window of the Interplanetary Space Age was greeted by successful firings from earth orbit of Venera 2 and Venera 3 on 12 and 16 November 1965, respectively. While Venera 2's main mission appears to have been a photographic flyby of Venus, Venera 3 was aimed at the heart of the planet itself.[25]

The Venera 2 probe weighed 963 kg and contained the now familiar cylindrical-type body, large parabolic antenna, propulsion unit, and two wing-like solar panels with hemispherical radiators of the thermal control system (Figure 4). The overall length was 3.6 m with a maximum width of 4 m.[26] Excluding the propulsion system the body consisted of two pressurized compartments: the large section called the orbital compartment and a smaller payload capsule. Inside Venera 2's and Venera 3's orbital compartments were located storage batteries, radio communications equipment, and spacecraft orientation sensing and control instruments. Radio transmission from the spacecraft to earth generally took place through the parabolic antenna with reception of earth transmissions through non-directional antennae. However, two-way communications could be conducted through the non-directional antennae in the event reorientation of the craft was not desired or possible.[27]

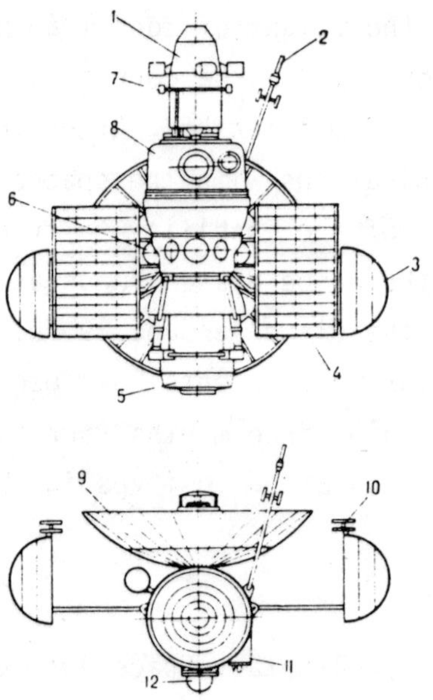

1—trajectory-correction engine unit; *2*—magnetometer probe; *3*—temperature-control system radiators; *4*—solar cell panels; *5*—special compartment; *6*—orientation system bottles; *7*—micro-engines of orientation system; *8*—orbital compartment; *9*—high-gain antenna; *10*—semi-directional antenna; *11*—continuous solar radiation sensor; *12*—precise stellar and solar orientation sensor.

Figure 4 Diagram of Venera 2

The second payload compartment of Venera 2 carried, along with some scientific measuring equipment, a phototelevision system.[28] With this apparatus Venera 2 was programmed to pass the sunlit side of Venus at a distance of less than 40,000 km with the hope of returning man's first close-up view of Venus.[29]

During the long flight to Venus the sister spacecraft were far from idle. Scientific instruments carried on Veneras 2 and 3 and used to examine the interplanetary medium included a three component magnetometer, cosmic ray gas-discharge counters and semiconductor detectors, micrometeor piezo-electric detectors (area = 1.5 m^2), ion traps for measuring flux and energy spectra of solar plasma, photon geiger counter, solar X-radiation detector, and receivers tuned to pick up cosmic radio emissions of wavelengths of 150 m, 1500 m and 15 km.[30]

Although equipped to perform any required midcourse corrections, Venera 2 was injected into an accurate enough trajectory to pass Venus at a distance of 24,000 km at 0552 hr Moscow time on 27 February 1966.[31]

The accompanying figure shows the relative positions of Venera 2, Venus, earth, and the sun at the time of bypass (Figures 5a and 5b). Unfortunately, as Venera 2 approached close to the planet an unexpected temperature increase was recorded and radio communications were disrupted.[32] Hence no real data on the Venerian environment was received after the otherwise successful three and one-half month flight.

The Venera 3 probe launched four days after Venera 2 was very similar in construction, weighing 969 kg, but was intended to perform an entirely different mission (impact on the surface of Venus).[33] While the orbital compartments of the spacecraft were virtually identical, the payload compartment of Venera 3 was a sphere of diameter 900 mm. The sphere was heavily insulated against the predicted intense heat of the Venerian atmosphere and the thermal energy generated during the process of deceleration in the atmosphere. Within the capsule were scientific instruments to measure the temperature and pressure of the Venerian atmosphere. Also included was a 70 mm diameter globe pennant of the earth and a Soviet medal to mark the first landing of a man-made object on another planet (Figure 6). In addition, the capsule was equipped with a parachute system to lengthen the descent time and possibly cushion the impact.[34]

Venera 3 underwent a course correction maneuver at 1804 hr on 26 December 1965 at a distance of 12.9 million km from earth.[35] Shortly before entering the atmosphere of Venus, Venera 3 was to eject its landing capsule. However, whether or not this actually happened is unclear. Like

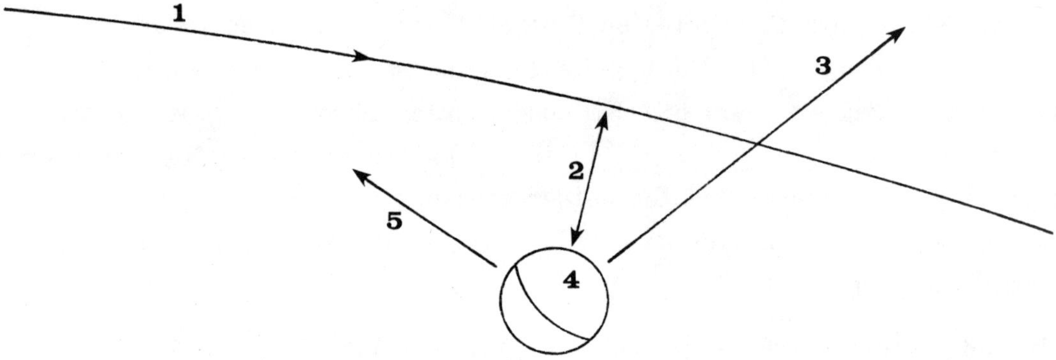

1. Trajectory of Venera 2
2. Minimum distance to Venus
3. Direction to the sun
4. Venus
5. Direction to the earth

Figure 5a Trajectory of Venera 2 during its encounter with Venus

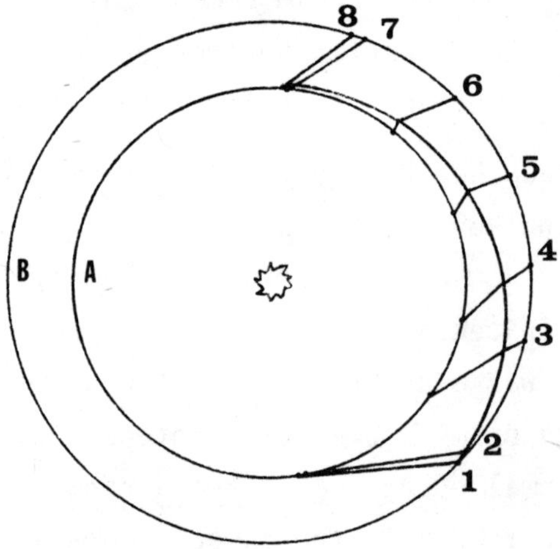

1.- 8.: Positions of earth-spacecraft-Venus during the flights

1. 12 Nov 1965, Venera 2 launch from earth
2. 16 Nov 1965, Venera 3 launch from earth
3. 10 Dec 1965
4. 26 Dec 1965
5. 15 Jan 1966
6. 4 Feb 1966
7. 27 Feb 1966, Venera 2 Venus flyby
8. 1 Mar 1966, Venera 3 Venus impact

A. Orbit of Venus B. Orbit of Earth

Figure 5b Trajectory of Veneras 2 and 3 from earth to Venus

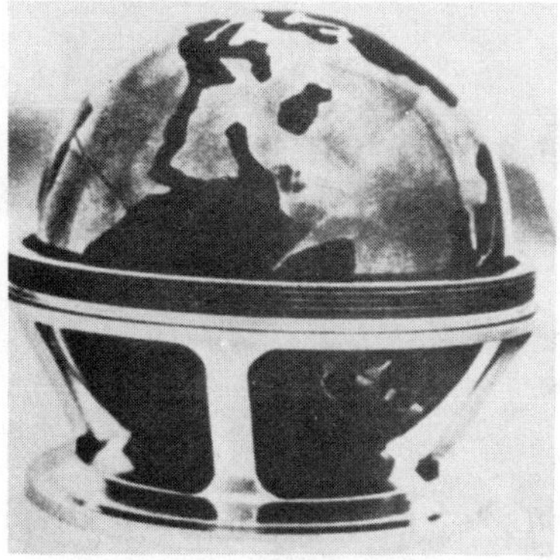

Figure 6 Earth globe carried by Venera 3 to the surface of Venus commemorating the first landing of a man-made object on another planet

its sister craft, Venera 3 experienced radio communication disruption as the probe approached the planet, apparently due to increasing temperatures.[36] No data were returned, but it is predicted that Venera 3 reached the surface of Venus on the night side near the terminator at 0956 hr Moscow time on 1 March 1966.[37] Figure 7 illustrates the calculated flight path. The capsule was reportedly sterilized to prevent contamination of Venus from earth bacteria. But the Soviet methods of sterilization and the fact that the unsterilized Venera 3 bus may also have landed on Venus drew some criticism from scientists outside the USSR.[38]

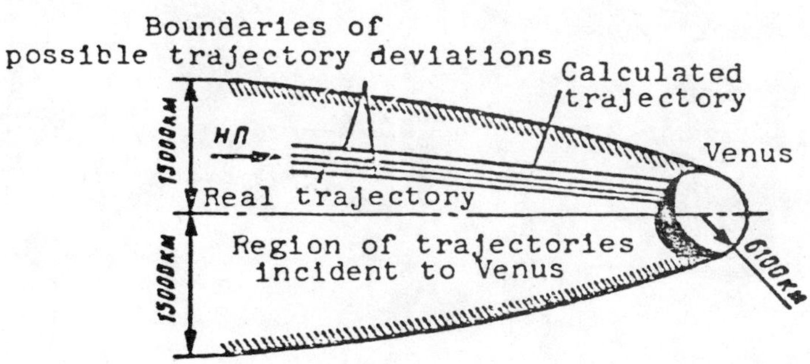

Figure 7 Intercept trajectory of Venera 3 on 1 March 1966

It is possible that a third probe was intended to follow Veneras 2 and 3 to the second planet from the sun. On 23 November 1965 Kosmos 96 was launched into an orbit very similar to that of Veneras 2 and 3. However, an orbital restart to inject the probe into the proper trajectory may have resulted in an explosion leaving eight pieces in orbit which decayed about sixteen days later.[39] In accordance with the Soviet practice of sending double probes to Venus, Kosmos 96 probably was identical to Venera 3. Venera 2 appears to have been a photo-reconnaissance probe left over from the Mars 1 and Zond 3 series.

VENERA 4, KOSMOS 167

The next launch window for Venus came during June of 1967. Again the Soviets were ready with what appears to have been a double launch. On 12 June Venera 4 entered a temporary low earth orbit and then was restarted on an intercept trajectory to Venus. Five days later Kosmos 167 entered a slightly higher orbit with a probable "Venera 5" payload. However the probe never left earth orbit and decayed 25 June 1967.[40]

During the 19 months since the launch of Veneras 2 and 3, Soviet space scientists had worked hard on improving the reliability and capabilities of the Venus probes. Similar in construction to her predecessor, the Venera 4 station weighed 1106 kg, including a 383 kg landing capsule (Figure 8).

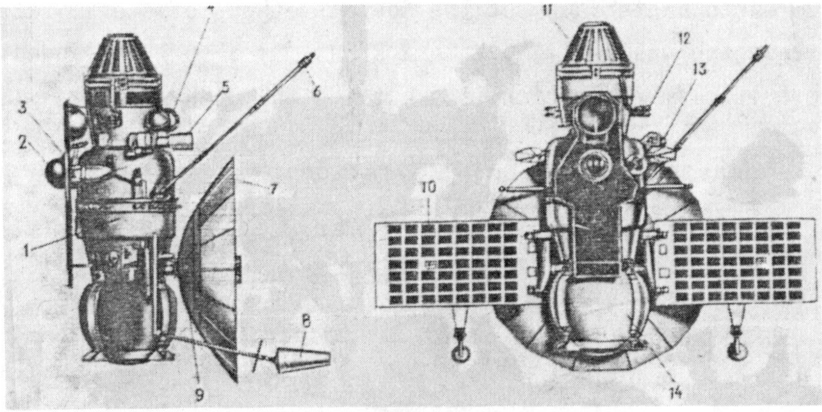

1—orbital compartment; 2—star sensor; 3—Sun sensor of continuous solar orientation; 4—gas bottles; 5—sensor of Sun—Earth orientation; 6—transducer and rod of magnetometer; 7—high-gain parabolic antenna; 8—low-gain antenna; 9—radiation panels of temperature control system; 10—panel of solar batteries; 11—vernier engine; 12—micro-rocket engine of astro-orientation system; 13—cosmic particles counter; 14—entry body.

Figure 8 Schematic diagram of the Venera 4 spacecraft

The length of the orbital bus sans parabolic antenna was about three and one-third meters. It measured a little over four meters across from the tip of one extended solar panel to the tip of the other. These measurements are almost identical with those of Veneras 2 and 3. The characteristic parabolic antenna had grown to approximately 2.3 m in diameter over the 2-m diameter antenna of Venera 1. As with previous craft there were also two non-directional antennae on the bus.[41] The solar panels fed and recharged the chemical storage batteries of the bus, and a final storage battery was located in the landing capsule to power the experiments and transmit their findings. This battery was not used during the flight to Venus and was only designed to supply the descent capsule with power for a minimum of 100 minutes after separation.[42] Thermal control for the orbital and landing compartments was by means of forced gas circulation, thus maintaining the compartments at a temperature between 15°C and 25°C.[43] In order to help monitor all systems and possibly correct malfunctions, an operating duplicate spacecraft was kept in an environmental chamber on earth during the Venera 4 flight.[44]

Scientific instruments carried on board the carrier bus included a boom magnetometer, cosmic ray counters, charged particle traps for ionospheric studies, and UV-radiation indicators to detect any hydrogen and oxygen. Magnetic field measurements were made in interplanetary space and in the vicinity of Venus beginning 16 August 1967. A study of the results of these measurements has been published.[45] The major findings were a lack of planetary magnetic field and corresponding radiation belts, but the presence of a shock wave in the solar plasma surrounding Venus was hinted at.[46] Although no atomic oxygen was detected, a weak hydrogen corona surrounding the planet at a distance of about 10,000 km was measured.[47]

A course correction was carried out on 29 July 1967 at a distance of 12 million km from earth using the liquid fuel correction engine.[48] At 0734 hr Moscow time 18 October 1967 as Venera 4 approached Venus, the descent capsule was detached from the carrier bus leaving the bus to be destroyed upon entering the atmosphere (Figure 9).

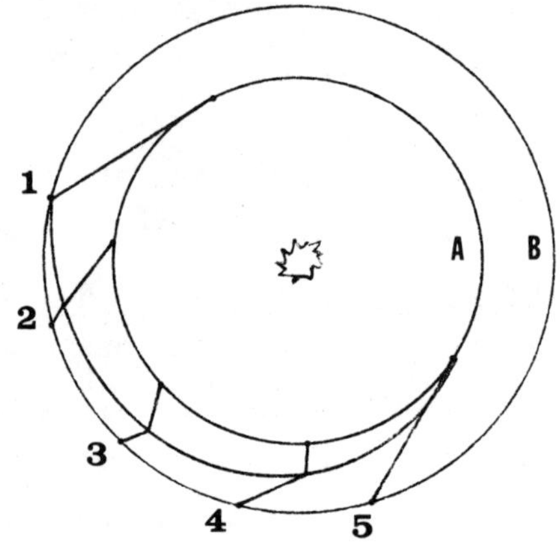

1. 12 June 1967, Venera 4 launched from earth
2. 12 July 1967
3. 12 August 1967
4. 18 September 1967
5. 18 October 1967, Venera 4 impact on Venus

A. Orbit of Venus
B. Orbit of the earth

Figure 9 Trajectory of Venera 4 between 12 June and 18 October 1967

The descent capsule resembled a sphere, one meter in diameter. It was covered on the outside with thick heat insulation concentrated at the forward end. The capsule was aerodynamically weighted so that the front end would bear the brunt of the heating during pre-parachute entry into the atmosphere. The capsule was made to withstand pressures of up to 20 kg/cm^2 and loads up to 300 g's. The parachute itself was designed to withstand temperatures of 400°C. In order to measure temperature, pressure, and atmospheric chemical composition two resistance thermometers, an aneroid barometer, an ionization densitometer, and 11 gas analyzer cartridges were used. Temperatures were measured using resistance-sensitive platinum wire in electrical bridge circuits, thus providing a temperature range of 0 - 400° C. An ionization-type densitometer was used employing β-active strontium-90 which was coated on the inner surface of a cylinder. The density was determined by the magnitude of a current between the cylinder and a central filament. The gas analyzers took two readings: the first immediately after opening the parachutes and the second 347 seconds later. Molecular nitrogen and oxygen, carbon dioxide, and water vapor were detectable with this device.[49]

In addition, a decimeter wavelength radio altimeter was used to determine the height of the descent capsule above the Venerian surface. The operation of this device and its subsequent ambiguous readings will be discussed later.

The descent capsule entered the Venerian atmosphere at an angle of 80° with a velocity of 10.7 km/sec 1500 km ± 500 km from the terminator near the equator on the night side of the planet.[50] Due to the intense aerodynamic heating, the temperature on the front of the capsule reached between 10,000 and 11,000°C. When the velocity of the spacecraft had been reduced to 300 m/sec, the upper part of the capsule was detached and a braking parachute was deployed. The main parachute was later deployed and the scientific instruments were turned on when the velocity had been reduced to 10 m/sec and a preset atmospheric pressure was recorded. This occurred at 0739 hr Moscow time with transmission from the capsule continuing for 93 minutes.

Initial readings from the radio altimeter indicated that the instruments were activated at an altitude of approximately 28 km and ceased upon reaching the surface. Maximum pressure and temperature readings of 18.5 kg/cm^2 and 270°C were recorded. However, the altimeter was of the frequency-modulated type, and sets of readings were possible. After comparision with data returned by the United States Mariner 5 spacecraft which passed Venus a day after Venera 4 arrived, it was determined that the instruments were activated at an altitude of 55 km. Maximum measurement capabilities of the barometric equipment occurred at an altitude of 36 km. At an altitude of 27 km the pressure exceeded the limits of the descent capsule and the upper cover of the equipment compartment collapsed, ceasing all communications.

Final interpretations of the data indicated a surface temperature and pressure of about 500°C and 75 atmospheres.[51]

VENERA 5, VENERA 6

Venera 4's descent through the atmosphere of Venus on 18 October 1967 supplied significant information on conditions present on the planet. However, the next launch window opportunity was scarcely more than a year away in January 1969. This did not allow adequate time to make substantial modifications needed if a new descent capsule was to survive all the way down to the surface of Venus. Therefore, when the next two probes of the veiled planet were launched, they were only slightly modified and better insulated versions of Venera 4. Their purpose was to amplify the data from Venera 4 and to return data below the 27 km altitude at which Venera 4 was crushed. The Soviet scientists knew these new probes would not be able to function under the conditions existing on the surface of the planet.[52]

At 0928 hr Moscow time on 5 January 1969, Venera 5 and its accompanying orbital platform were launched into earth orbit. One hour and 19 minutes later, Venera 5 was on its way to Venus. Five days later, Venera 6 was also successfully launched and placed on a Venus intercept trajectory. Both stations weighed in at a record 1130 kg.[53]

The orbital buses of Venera 5 and Venera 6 were virtually identical with that of Venera 4. The 2.33-m parabolic antenna was used only at extreme distances from the earth for high-rate transmissions. The two solar panels covered an area of 2.5 m^2 and constantly recharged the chemical batteries of the orbital compartment. As the probes moved closer to the sun, the solar flux increased, necessitating control of individual regions of the solar panels to prevent overcharging the batteries.[54]

The attitude-control system was broken up into three major areas: solar orientation, orientation of the parabolic antenna toward earth, and orientation of spacecraft during midcourse trajectory maneuvers. A null-seeking solar sensor activated microengines to maintain the solar panels' orientation toward the sun. In addition, to minimize attitude drifting, the spacecraft was rotated around the sun-spacecraft axis in a manner similar to the early manned Soyuz spacecraft. During radio communications via the parabolic antenna, the solar-orientation program was modified to point the antenna beam toward earth. Upon completion of communications, the station resumed its original orientation. Venera 5 underwent a midcourse maneuver on 14 March 1969 at a distance of 15.5 million km from earth, whereas Venera 6 performed a similar maneuver on 16 March at a distance of 15.7 million km. During these engine burns the orientation of the probes was determined using the sun and the star Sirius as reference points.[55]

The 1-m diameter descent capsules of Veneras 5 and 6 were only slightly improved over that of Venera 4. The weight of each capsule was 405 kg, an increase of 22 kg. The new capsules could withstand pressures up to 27 kg/cm^2 and loads up to 450 g's, versus 20 kg/cm^2 and 300 g's for Venera 4. Most importantly, the parachute size was reduced to approximately one-third the size of Venera 4's chute and could withstand temperatures of 500°C. This, coupled with a slightly higher entrance velocity into the atmosphere (11.2 km/sec) and a lower preset velocity for deployment of the parachutes (210 m/sec), would hopefully allow the probes to descend further into the atmosphere before being destroyed. An improved radio altimeter was also installed to avoid the problems encountered by Venera 4.[56]

The Venera 5 descent capsule was ejected from the orbital bus at a distance of 37,000 km from Venus. At 0901 hr on 16 May 1969 the probe entered the atmosphere on the dark side of the planet 2700 km from the terminator at an angle of 62-65° to the horizontal at a velocity of 11.2 km/sec. Although the orbital bus disintergrated upon reaching the dense atmosphere, the descent capsule rapidly decelerated, causing the temperature on the front of the capsule to reach 11,000°C. When the velocity had been reduced to 210 m/sec, the drogue chute was deployed, followed by the main chute. After the main parachute was out, communications were reestablished with earth (at 0902 hr), the radio altimeter antennae were extended, and the other scientific equipment began collecting data on temperature, pressure, density, and composition of the Venerian environment. Communications lasted for a period of 53 minutes at which time the pressure exceeded 27 kg/cm^2, crushing the probe at an altitude of 24-26 km after collecting data through a distance of 36 km. During the descent the interior temperature rose from 13°C to 28°C.[57]

Almost exactly one day later at 0905 hr on 17 May the Venera 6 descent capsule entered the Venerian atmosphere, having been ejected from its carrier at a distance of 25,000 km from the planet. It, too, entered the atmosphere on the night side at an angle of 62-65° approximately 300 km from the entry point of Venera 5. For 51 minutes and a distance of 38 km, Venera 6 returned data. Communications ceased abruptly at an estimated altitude of 10-12 km. Since both probes ceased operating at the same pressures, it is assumed that the variance in altitudes between the two craft was probably due to Venera 6's landing on a mountain or high plateau. Recent discoveries concerning the Venerian terrain at least make this a possibility.[58]

As with Venera 4 before them, Venera 5 and Venera 6 used gas analyzers to measure amounts of oxygen, nitrogen, carbon dioxide, and water vapor in the atmosphere. Two samples were made by each craft at different altitudes. Temperature and pressure measurements were made by the same techniques described with Venera 4. One modification provided Veneras 5 and 6 with a different type of densitometer. These new instruments operated on the principle of a tuning fork made of steel plates 12 mm wide and 0.6 mm thick. An electrical signal proportional to the amplitude of the vibration recorded changes due to density variations. Additionally, photoelectric sensors were carried to measure illumination as small as 0.5 w/m^2. No readings were recorded with this latter sensor with the exception of one measurement of 25 w/m^2 from Venera 5, four minutes before the capsule was destroyed.[59]

Analysis of the returned data indicates surface conditions of 140 atmospheres and 530°C from Venera 5 and a pressure of 60 atmospheres and a temperature of 400°C for Venera 6. The atmosphere was found to be mainly carbon dioxide (93-97%) with traces of nitrogen and inert gases (2-5%) and of oxygen (less than 0.4%).[60] These results closely confirmed those of Venera 4, and proved that Venus was indeed an even more inhospitable planet than most of her detractors had imagined.

Now with the possibility of man ever landing on Venus all but waived, few would have questioned diverting the time and expense of exploration to perhaps more promising areas, perhaps a more concentrated effort on Mars or the moons of Jupiter. Soviet scientists, though, were determined not to abandon their adopted planet, but instead to probe more deeply the secrets hidden beneath those eternal clouds. To do this would require major modifications of the landing apparatus. They had until August 1970.

VENERA 7 AND KOSMOS 359

The 17th of August 1970 marked the 18th attempt of the USSR to investigate Venus and its seventh attempt to return data from her surface. On that day Venera 7 was placed into a parking orbit, and one hour and 21 minutes later was fired toward Venus. As with Venera 5 in 1969 and Venera 4 in 1967, five days later a second satellite was placed in earth orbit. What would have been Venera 8 was ejected from the orbiting platform but apparently experienced some engine malfunction, moving only to a more elliptical orbit. The satellite was subsequently designated Kosmos 359.[61]

Not unexpectedly there was a great similarity between Venera 7 and her previous sister craft. The major design feature changes were to be found in the landing capsule itself. The entire Venera 7 station weighed 1180 kg with the landing capsule now accounting for about 500 kg. The descent module was now capable of withstanding pressures up to 180 kg/cm^2 and a temperature of 540°C for one to one and a half hours. The hermetically sealed compartment containing the scientific instruments and communication devices was now a perfect sphere in order better to withstand high pressures. The outer insulation and shock absorbers gave it a more egg-shape appearance. The shock absorber was new since, for the first time, Soviet scientists actually expected the probe to survive all the way down to the surface.[62]

Unlike previous probes Venera 7 required two midcourse corrections to insure proper entry into the Venerian atmosphere. These occurred on 2 October and 17 November 1970 when 17 million km and 31 million km from earth, respectively. On 12 December at a distance of 1.3 million km from Venus, the chemical batteries of the descent capsule were ordered charged using the solar panels of the carrier bus. At this same time a new procedure was initiated: the descent capsule temperature was reduced to -8°C. Survival of the probe, which was now largely dependent on how soon the intense temperature of the surroundings penetrated the probe, was thus prolonged.[63]

To insure separation of the landing capsule from the carrier bus, several independent procedures were available. The command to separate might come directly from earth, from the on-board programmer, from the overload sensor which would trigger separation upon experiencing high g loads while entering the atmosphere, and from a sensor which was activated when communications were broken with earth by the disturbance of the craft's orientation upon entering the atmosphere. As a last resort, if all the above failed, the bands attaching the capsule to the station were designed to burn and break during initial entry into atmosphere. At approximately 0800 hr on 15 December, the landing capsule was separated as the station entered the atmosphere at an altitude of 135 km.[64]

The landing module entered with a velocity of 11.5 km /sec at an angle between 60 and 70 degrees. Aerodynamic drag reduced the velocity to about 200 m/sec, and at an altitude of 60 km the parachute system was activated. Within about 35 minutes the landing capsule had come to rest on the Venerian surface, 2000 km from the morning terminator. This much quicker descent time than previous Veneras was made possible by the significant reduction in the size of the parachutes. At the moment of impact all communications appeared to have ceased. This would have represented another bitter disappointment to the Soviet scientists who had been trying to conquer Venus for the past ten years. Luckily, the Soviet tracking network continued to record what appeared to be unintelligible noise. After much computer reprocessing, 23 minutes of valuable data were extracted. Apparently the capsule had landed in such a way as to prevent a clear transmission to earth. The revived signals were at a strength of 0.01 of the power transmitted prior to landing .[65]

The temperature sensors on board the capsule had a range of from 25°C to 540°C and pressures could be measured between 0.5 and 150 atmospheres. From the readings received while the probe was on the Venerian surface, a temperature equal to 475° ± 20° C and a pressure equal to 90 ± 15 atmospheres were calculated.[66]

VENERA 8 AND KOSMOS 482

The Soviet Union had not let a single Venus launch window pass without launching or attempting to launch two or more probes since the beginning of the Planetary Space Age. After the successful flight of Venera 7, the next window fell in March 1972. As if on key, Venera 8 was launched and placed on a Venerian trajectory on 27 March 1972. Four days later, what was to have been Venera 9 was put into an earth orbit of 483 km by 204 km. The intended probe apparently misfired (as had Kosmos 359 in August 1970), sending it into only a more elliptical orbit of 9812 km by 210 km. The craft was subsequently designated Kosmos 482.[67] It has been reported that Kosmos 482 was to land on the dark side of the planet while Venera 8 landed on the day side.[68]

Again the basic spacecraft closely resembled her sisters. The orbital bus was virtually unchanged with the total weight of the station amounting to 1184 kg including 495 kg for the lander. However, by incorporating the results of the Venera 7 experiments, the descent capsule was substantially modified. Since the temperature and pressure at the Venerian surface had now been firmly established, the overdesign of the Venera 7 capsule could be reduced. The capsule of Venera 8 was built to withstand pressures of 105 atmospheres, versus 180 atmospheres for Venera 7. The depleted weight of structural components was replaced by additional scientific instruments and heat-absorbing insulation. The parachute was also strengthened with respect to that carried on Venera 7. To insure that the transmission problems after the landing of Venera 7 did not reoccur, the Venera 8 capsule ejected on to the planet's surface a special second antenna after landing. Radio communications could be maintained using either antenna.[69]

In addition to the standard measurements of temperature, pressure, and height several new instruments were added. Since Venera 8 would be the first man-made object to probe the day side of the planet, a series of cadmium sulfide photo-resisters was used to measure the light flux through the atmosphere and on the surface. The amount of ammonia in the atmosphere was also to be measured by recording the change in color of ammonia reactive tetrabromphenol sulfoftalein with photo-resisters. Horizontal wind speeds at different altitudes were to be recorded. Finally, several radioactive elements (thorium, uranium, and potassium) of the Venerian soil were to be searched for by examining the spectra of the soil gamma radiation. These findings could then be used to postulate the nature of the planet's surface.[70]

The trip to Venus was uneventful and required only one course correction which was performed on 6 April 1972. Fifty-three minutes before entry into the atmosphere, the descent capsule was separated from the orbital bus. Employing the same procedure as first used with Venera 7, the descent capsule batteries had been charged several days earlier and the interior of the capsule had been lowered to -15°C. The new high heat-capacity material surrounding the instrument compartment within the capsule would now be able to absorb a great deal of heat while on the surface of the planet, substantially prolonging the probe's life. The descent capsule entered the atmosphere at an angle of 77 degrees with a velocity of 11.6 km/sec at 1137 hr on 22 July 1972. In accordance with the ever-decreasing descent times of the Venera spacecraft through the atmosphere, the Venera 8 lander had decelerated to 250 m/sec and deployed its braking parachute within 18 seconds of atmospheric entry. For the first time, the main parachute was initially reefed (by 70%) and at an altitude of 30 km the chute was fully opened, permitting landing to take place at 1232 hr some 500 km on the sunlit side of the morning terminator. All systems continued to function for an additional 50 minutes on the surface (Figure 10).[71]

Interpretation of the data received indicated that the temperature and pressure at the Venera 8 landing site were 470 ± 8° C and 90 ± 1.5 kg/cm^2. The amount of ammonia in the atmosphere measured at altitudes of 46 and 33 km was calculated to be between 0.1 and 0.01 percent. Wind speeds ranged from 100 m/sec above 48 km to 40-70 m/sec between 42 and 48 km to less than 1 m/sec below 10 km. The much awaited results on light flux measurements indicated that a sharp change in illumination occurred at an altitude of between 30 and 35 km. It has subsequently been predicted that

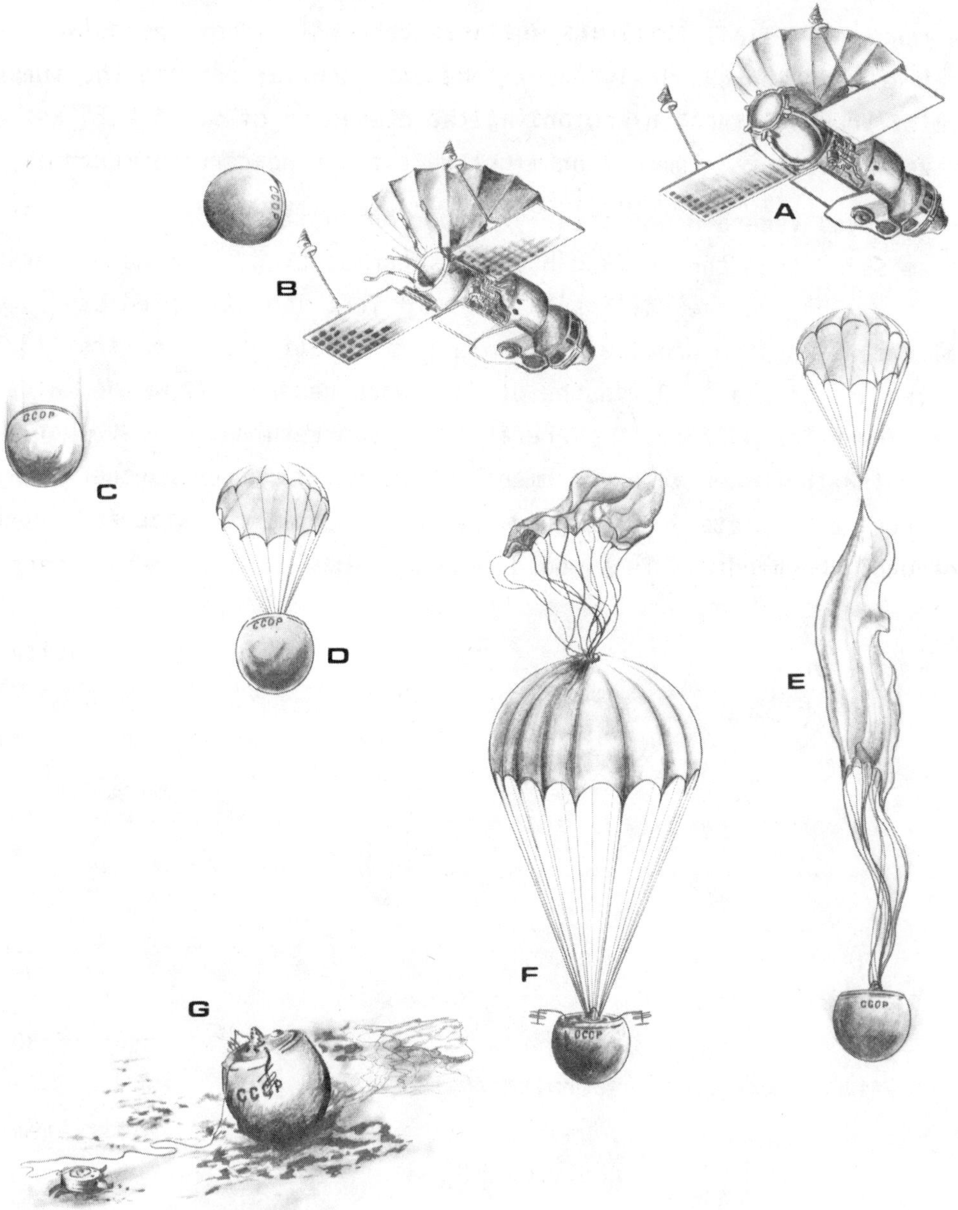

A. Venera 8 as it approaches Venus
B. Separation of descent capsule from carrier bus
C. Entry into Venerian atmosphere
D. Deployment of drogue parachute
E. Deployment of main parachute
F. Activation of scientific instruments
G. Landing on the Venerian surface

Figure 10 Descent sequence of Venera 8 (R. Escarcega)

the range of visibility at the surface where only 1% of the solar flux penetrates is approximately one kilometer. However, during the operational time of Venera 8 the sun was only at an elevation of 5.5 ± 2.5° above the horizon. Greater illumination might well occur nearer the Venerian noon.[72]

The improved Venera 8 had clearly outperformed its predecessors. Time of transmission from the surface had more than doubled, 50 minutes versus 23 minutes. The new thermal insulation had functioned as predicted and the dual antenna system provided reliable communications. The first 13.3 minutes and the last 30 minutes of data were returned from the main antenna on the Venerian surface.[73] Veneras 4 - 8 had returned much valuable data concerning the Venerian environment. However, to probe further the secrets of Venus, a completely redesigned craft of much larger size and capabilities would be needed. This would require time.

Chapter 2

SECOND-GENERATION SPACECRAFT

VENERA 9 AND VENERA 10

October-November of 1973 approached and passed without any attempts to launch a Soviet Venus space probe. This represented the very first opportunity in which no attempts were made, since the first launchings in February of 1961. When the next window came in June of 1975, the reason for the brief hiatus was clear. On 8 June a massive spacecraft designated Venera 9 was first put into earth orbit and later fired into a Venerian trajectory. At a total weight of 4936 kg, Venera 9 was over four times heavier than any previous Venus probe. Four days later, the second of the usual Soviet dual-probe program was launched. Venera 10, weighing 5033 kg, was also successfully placed on an intercept course with Venus.[74] Both probes, representing the second generation of Venus explorers, were launched by the so-called Proton booster. The largest capacity operational booster in the Soviet inventory, the Proton rocket has been used to launch the Salyut manned space stations (beginning in 1971), the Zond man-oriented lunar flights (beginning in 1968), the third generation Luna probes (beginning in 1969), and the second-generation Mars probes (beginning in 1971).

The new Venera probes bore little resemblance to their predecessors (Figure 11). The craft were divided into two major components: an orbiter and a lander. No longer was there to be an expendable carrier bus which returned relatively little data, but one consisting of up to 65% of the total spacecraft weight. The main mission of the orbiters, which were to become the first two satellites of Venus, was to relay scientific data collected by the landers to earth. In addition, the orbiters, which weighed approximately 2300 kg after insertion into orbit, were equipped to photograph the Venerian cloud cover in UV light in strips 1200 km wide with the French-built 3500 Å range UV photometers. Other instruments carried on board to investigate the clouds included: a 4000-7000 Å range photo-polarimeter (cloud brightness), a 1.5 - 3 micron range infrared spectrometer (reflectivity), and a 8 - 30 micron range infrared radiometer (temperature). To monitor interplanetary and planetary magnetic fields were a magnetometer and charged particle traps. Lastly, it has been reported that the orbiters carried a photo system to return images of the planet.[75]

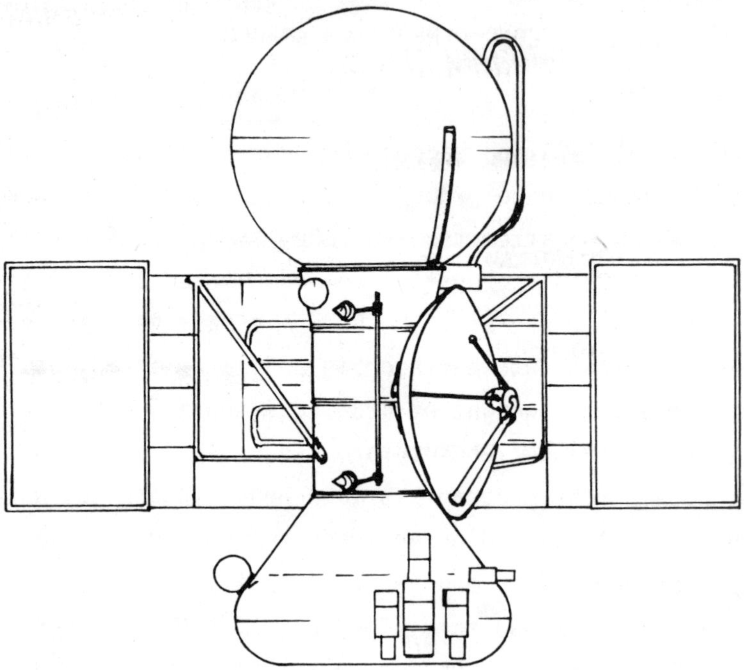

Figure 11 Sketch of the second-generation Venera explorers (R.F. Gibbons)

The new Venera lander had undergone a more radical change than the carrier bus. Although the instruments were encased in a sphere of about 2.4 m in diameter while attached to the mother craft, the actual lander was of an entirely new design (Figure 12). As with the previous Venera descent capsules, Veneras 9 and 10 employed the aerodynamic braking of the dense Venerian atmosphere to decelerate the lander without the need of added retro-rockets common to Lunar and Martian probes. After the main parachutes had completed their task, they were detached at an altitude of 50 km. The landers then "fell" the remaining distance to the surface, retarded only by the upper circular air brake (the air brake served the dual purpose of acting as reflector for the upper fixed antenna). Upon impacting the surface at a velocity of approximately 7 m/sec, a series

Figure 12 Venera 9/10 lander undergoing tests on earth (TASS from SOVFOTO)

of shock absorbers and a deformable ring around the bottom of the probe absorbed the landing energy.[76]

The lander contained a much larger array of instruments than its forerunners. The most intriguing piece of new equipment was a panoramic telephotometer. For the first time in history, man was about to see the surface of Venus, heretofore hidden beneath an immense permanent cloud cover. Located underneath the air brake were several portholes which, when uncovered, permitted a 140 - 160 degree panoramic view of the surface. Spotlights of 10,000 lux were positioned beneath these windows to illuminate the predicted dark surroundings. Other photometers in the range of 0.8 and 1 micron were available to investigate light penetration and atmospheric composition. As with Venera 8 a gamma-ray spectrometer, mass spectrometers, an anemometer, and temperature and pressure sensors were on board. In addition, a radiation densitometer was placed on the surface after landing to determine soil density.[77]

On 20 October 1975 the Venera 9 lander capsule separated from the orbiter. Similar to Veneras 7 and 8, the Venera 9 lander interior was cooled to -10°C while at the same time its exterior temperature was reduced to -100°C. Two days later at 0658 hr the capsule entered the atmosphere at a much shallower angle than previous probes (about 20.5 degrees to the horizontal). Its initial velocity of 10.7 km/sec was quickly reduced to 250 m/sec at an altitude of about 64 km where the upper clamshell cover was ejected and the first braking parachute was released. A second drag chute, then three main parachutes were successively deployed, and the lower clamshell cover was discarded. Twenty minutes later just below the cloud cover at an altitude of 50 km, the main parachutes were jettisoned. Even at this great height the dense atmosphere of Venus and the Venera airbrake prevented Venera 9 from accelerating to its destruction. The lander came to rest on the day side of the planet at 0813 hr (Figure 13).[78]

The approximately two-meter high craft had landed at a tilt of between 15 and 25 degrees. During its 53-minute lifetime on the Venerian surface, the Venera 9 lander took and transmitted the first photo ever taken on the planet (Figure 14). Surprisingly enough the probe floodlights were not even needed. The solar illumination was measured at 14,000 lux, equivalent to a cloudy day on earth. The historic picture clearly showed sharp flat rocks surrounding the craft.[79]

The Venera 10 spacecraft followed an almost identical pattern. The lander separated on 23 October and entered the atmosphere at an angle of 23 degrees on 25 October at 0402 hr. By 0517 hr it had successfully landed slightly more than 2000 km from the Venera 9 site. Its 65-minute life of transmitting scientific data established a new record and was more than double the 30-minute design life. A picture of the Venera 10 surroundings revealed a much smoother surface of what appeared to be weathered rocks (Figure 15).[80]

While the two landers began their descent, the orbiters of Venera 9 and 10 entered orbit about the planet from the side opposite to the landers' entry to aid in communications during the short transmission times of the landers. Venera 9 entered an orbit of 1300 - 112,000 km with orbital inclination of 34°10'. Venera 10's parameters were 1400 - 114,000 km and 29°30'.[81]

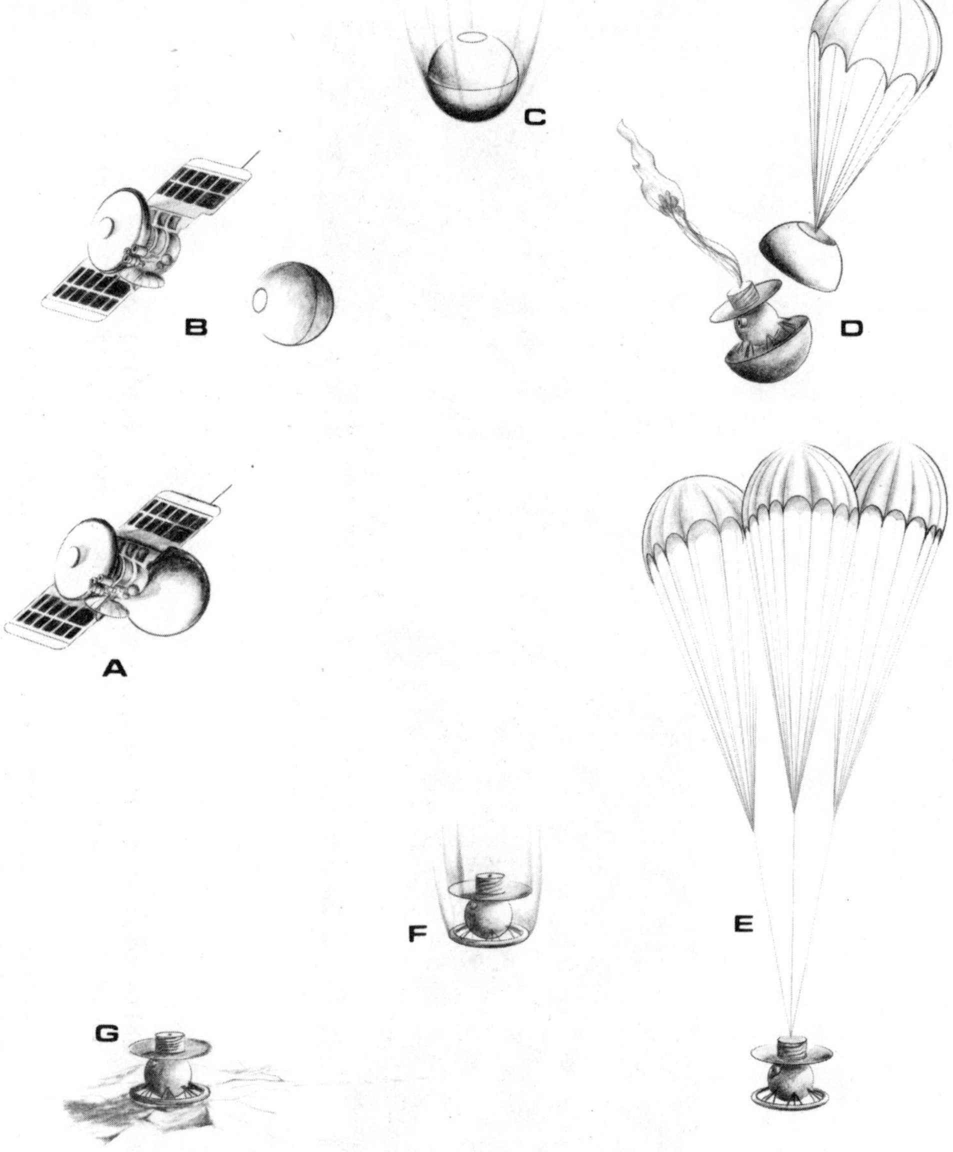

A. Approach of Venera to the planet Venus
B. Separation of the landing capsule from the orbital bus. The bus later fires its engine to enter into orbit about Venus.
C. Entry into the Venerian atmosphere
D. Separation of braking parachute and upper portion of lander cover
E. Deployment of main parachutes
F. Free fall of lander through Venerian atmosphere
G. Venera lander on the surface of Venus

Figure 13 Descent sequence of Venera 9/10 (R. Escarcega)

White segment on the bottom is part of the landing system
Arrow on the lower right shows density meter lying on the surface

Figure 14 Historic first photograph of the Venerian surface returned by Venera 9 on 22 October 1975 (NOVOSTI from SOVFOTO)

Figure 15 Photograph returned by the Venera 10 lander on 25 October 1975 (NOVOSTI from SOVFOTO)

Preliminary analysis indicates that both landers came to rest at an altitude of 1.5 to 2 km above "sea level". Venera 9 appeared to have landed at the base of a hill while Venera 10 rested on an older plateau. The Soviets later revealed that the cessation of data from the landers was not a result of their own demise, but due to the respective orbiters which relay the transmissions from the landers to earth moving out of range. Actual survival time of the landers on the surface might have been ninety minutes. Pressures between 85 and 92 kg/cm^2 and temperatures between 455°C and 465°C were recorded by the probes with the wind velocities at both sites below 4 m/sec. Both gamma-ray spectrometers and radiation densitometers indicated that the surface layer resembled basalt rather than the granite as first thought by the readings of Venera 8.[82]

The Venera 9 and 10 probes were unqualified successes. The landers returned the first pictures of the surface along with much additional information not recorded by either Venera 7 or Venera 8. Also, the orbiters continued to relay significant amounts of information about the planet and near-interplanetary space for many months.

VENERA 11 AND VENERA 12

The highly successful missions of Veneras 9 and 10 must logically have called for a reevaluation of the Venus exploration program by Soviet program managers. The information gathered from the two orbiters and two landers in 1975 along with the atmospheric and surface data compiled by Veneras 4 - 8 verified the harsh environment of Venus, and illustrated the great difficulty of employing more sophisticated surface probes. However, further opportunities for Venus orbiters and upper atmospheric balloons existed if the funds for continued Venus exploration were not diverted to more promising celestial bodies.

During the summer of 1976 Soviet spokesmen acknowledged that no Venera spacecraft were to be launched in the upcoming winter 1976-1977 launch window, marking only the second time in fifteen years that an attempt would not be made to explore Venus. The reason for this decision was attributed to the time required to fully analyze the data returned by Venera 9 and Venera 10 and to develop improved landing craft. Mention was also made of preliminary discussions with the French to manufacture instrument packages designed to float in the Venerian atmosphere for several hours or days.[83]

By the end of 1977 more definite plans for the future investigation of Venus had been revealed. Two new Venera probes were to be launched at the 1978 launch window to rendezvous with Venus in December of that year at the same time that the United States was to attempt its very first Venus atmospheric and orbiter missions.[84] However, a much more ambitious Soviet-French undertaking was scheduled for 1983.

According to the plan, shortly before entering a Venerian orbit, a Soviet spacecraft will release a French payload into the atmosphere. After opening at an altitude of 60 km, two balloons will fall an additional 11 km before fully inflating to their 8-9 m maximum diameter. The balloons are expected to continue floating in the Venerian atmosphere for perhaps up to six days.[85]

While this later project was still in the early developmental stages, the USSR fulfilled her promise to send two Venera probes during the 1978 opportunity. On 9 and 14 September 1978 near the end of the launch window Veneras 11 and 12 were inserted into earth orbits of 205 km by 177 km at 51.5° before beginning their three and one-half month treks to Venus. Unlike the Venera 9 and 10 spacecraft Venera 11 and 12 which are estimated to have weighed between 4500 kg and 5000 kg each consisted only of a lander craft and a flyby bus due to the higher energy requirements in 1978 than in 1975.[86] A similar situation and solution had occurred during the 1971 and 1973 Mars attempts. One advantage of the lander/flyby bus combination over the earlier lander/orbiter configuration was the increased contact time between the landers and the mother craft. Surface data for a period of up to two hours was hoped for.

Instrumentation on board Venera 11 and 12 had been expanded since the last Venus probes three years earlier. A new 10-kg gas chromograph was designed to measure more accurately the composition of the Venerian atmosphere, including the presence of carbon dioxide, carbon monoxide, nitrogen, and isotopes of argon. In addition to instruments for analyzing the spectrum of scattered solar radiation and for detecting soil composition, the descent capsules were equipped with a device named Groza for the purpose of detecting atmospheric electrical discharges. Telephotometers were also carried to return new panoramas of the scorched surface. The flyby buses along with their role of communications relay between the landers and earth were capable of examining the solar wind, cosmic rays, and a wide range of electromagnetic radiation.[87]

Although launched last, the Venera 12 complex arrived first at Venus after course corrections on 21 September and 14 December. On 19 December, just two days before the encounter, the subcooled landing capsule was separated from the carrier bus. Finally, on 21 December the spherical payload slammed into the dense Venerian atmosphere with a velocity of 11.2 km/sec. The ensuing descent procedure was virtually identical to that of Veneras 9 and 10 with the circular air brake taking over from the parachute system at an altitude of 40 km. Touchdown came at 0630 hr Moscow time after a descent of about one hour.

Early returns from Venera 12 indicated that surface conditions at the landing site were approximately 90 atmospheres pressure and 460°C temperature. Data continued to be relayed to earth for 110 minutes until the Venera 12 bus which had passed the planet by 35,000 km was out of sight. Apparently no television pictures of the surface were received.

Venera 11 arrived on the scene on 25 December after performing course corrections on 16 September and 17 December. The Venera 11 landing capsule was also ejected two days prior to encounter and entered the atmosphere at 11.2 km/sec. By 0624 hr on Christmas Day, Venera 11 had successfully endured the searing descent and came to rest on the Venerian surface some 800 km away from Venera 12.

The 95 minutes of data from the surface were again prematurely ended as the Venera 11 bus flew out of range, following a similar 35,000 km flyby. For presently unknown reasons, like Venera 12 four days earlier no photographs of the Venera 11 site were returned.

One of the major findings of Veneras 11 and 12 was the presence of violent thunder activity. One thunder clap recorded after the landing of Venera 12 lasted for fifteen minutes. Also of interest was the detection of the ratio of argon 36 to argon 40 some 200 to 300 times that found on earth, and the discovery of carbon monoxide at lower altitudes.[88] All these data if confirmed will now have to be incorporated into new theories of planetary origin and evolution which may one day help us to better understand our own world.

Apart from the already announced 1983 Soviet-French mission, the future of Venus exploration is not clear. Although the Soviets have talked of placing a radar mapper in orbit about Venus, enthusiasm for the project does not appear high. In the near term, improvements in the lander communications equipment might allow direct contact with earth, and hence provide longer data-return periods.

In the decades to come the Soviets have expressed the desire to retrieve a soil sample from the surface. An alternate mission profile would employ a roving type vehicle similar to Lunokhod to survey an extended area. The extreme conditions present on the Venerian surface, however, make the feasibility of such a rover uncertain due to the expected short lifetime of the craft. Both projects will undoubtably require spacecraft larger than is possible with current Soviet rocket boosters.

In the meanwhile, Soviet interest in Venus remains high, and innovations in Venus exploration are almost certain to appear in the future.

REFERENCES AND NOTES

1. Gatland, K., *Spacecraft and Boosters*, p. 10

2. *TRW Space Log, 1975*, p. 40

3. *TRW Space Log, 1975*, p. 40; *Soviet Space Programs, 1971-1975*, p. 85

4. *Space World;* Jan. 1975, p. 11; *Soviet Space Programs, 1971-1975*, p. 85; *Aviation Week and Space Technology*, 17 Sep. 1962

5. *Soviet Space Programs, 1971-1975*, p. 554

6. Gatland, K., *Spacecraft and Boosters*, p. 10

7. *TRW Space Log, 1975*, p. 40; *Soviet Space Programs, 1971-1975*, p. 554

8. These three included the two Mars failures of Oct. 1960 as well as Sputnik 7.

9. Skuridin, G.A., *Mastery of Outer Space in the USSR, 1957-1967*, pp. 1107.1110

10. Alekeseyev, V. and Minchin, S., *Venus Reveals Secrets*, p. 30; Ponomarev, A.N., *The Years of the Space Era*, p. 196; Gatland, K., *Robot Explorers*, pp.168-169

11. Petrovich, G.V., *The Soviet Encyclopedia of Space Flight*, p. 487

12. Skuridin, G.A., *op. cit.*, p. 1125; Aleksandrov, S.G. and Federov, R.Ye., *Soviet Satellites and Space Ships*, pp. 166-178

13. Skuridin, G.A., *op. cit.*, pp. 1125-1127

14. Skuridin, G.A., *op. cit.*, p. 1126

15. Wukelic, G.E., *Handbook of Soviet Space-Science Research*, pp. 169, 285; Blagonravov, A.A., et al., *USSR Achievements in Space Research*, p. 610

16. *Science*, 24 Mar. 1967, p. 1509

17. Skuridin, G.A., *op. cit.*, p. 1128

18. *TRW Space Log, 1975*, p. 44; *Science News*, 22 July 1967; *Aviation Week and Space Technology*, 10 Sep. and 17 Sep. 1962; Gatland, K., *op. cit.*, p. 47

19. *TRW Space Log, 1975*, p. 47

20. *Soviet Space Programs, 1971-1975*, p. 87

21. *TRW Space Log, 1975*, p. 48; *Aviation Week and Space Technology*, 30 Mar. 1964; *New York Times*, 1 Apr. 1964, pp. 1, 5; *Astronautics and Aeronautics*, Apr. 1964, p. 88

22. Wukelic, G.E., *op. cit.*, p. 72; Blagonravov, A.A., et al., *op. cit.*, p. 610; Petrovich, G.V., *op. cit.*, p. 509

23. *Soviet Space Programs, 1971-1975*, p. 87; *Science*, 24 Mar. 1967, p. 1509

24. Avdyushin, S.I., et al., *Ionizing Radiation Intensity Based on Measurements on Zond 1*

25. *Spaceflight*, May 1966, p. 163 and Aug. 1966, pp. 297-298 and Nov. 1966, pp. 392-393

26. Blagonravov, A.A., et al., *op. cit.*, p. 610

27. Skuridin, G.A., *op. cit.*, pp. 1205-1207

28. Petrovich, G.V., *op. cit.*, p. 487; Skuridin, G.A., *op. cit.*, p. 1203; Blagonravov, A.A., et al., *op. cit.*, p. 610; Wukelic, G.E., *op. cit.*, p. 307; Shelton, W., *Jane's All the World's Aircraft, 1966-1967*

29. *Astronautics and Aeronautics*, Oct. 1967, p. 75

30. Skuridin, G.A., *op. cit.*, p. 1222; Blagonravov, A.A., et al., *op. cit.*, p. 611. (Results of some of these experiments can be found in Wukelic, G.E., *op. cit.*)

31. Skuridin, G.A., *op. cit.*, p. 1199; Alekeseyev, V. and Minchin, S., *op. cit.*, p. 32

32. Alekeseyev, V. and Minchin, S., *op. cit.*, p. 33

33. Some reports initially indicated that Venera 3 would be a flyby on on the opposite side of Venus. *Science*, 24 Mar. 1967, pp. 1507-1508

34. Skuridin, G.A., *op. cit.*, pp. 1197, 1204; *Science*, 24 Mar. 1967, p. 1508

35. Skuridin, G.A., *op. cit.*, p. 1215; Blagonravov, A.A., et al., *op. cit.*, p. 611

36. Alekeseyev, V. and Minchin, S., *op. cit.*, p. 33; Gurney, G. and C.; *Cosmonauts in Orbit*, p. 159

37. Skuridin, G.A., *op. cit.*, pp. 1198-1199; *Ten Years of Space Exploration*, p. 22; *McGraw-Hill Encyclopedia of Space*, p. 715. (Some results of Venera 3 experiments conducted during its flight can be found in Wukelic, G.E., *op. cit.*)

38. *Science*, 24 Mar 1967, pp. 1505-1510. This article contains some details of sterilization techniques, but assumes that the Soviets intended to deflect the Venera orbital buses as well as the final rocket stages away from the planet after the descent capsule release.

39. Shelton, W., *op. cit.*, p. 231; *Soviet Space Programs, 1971-1975*, p. 562; *Science News*, 22 July 1967

40. *TRW Space Log, 1975*, p. 62; *Soviet Space Programs, 1971-1975*, pp. 90, 657

41. Skuridin, G.A., *op. cit.*, p. 1231a; *Pravda*, 22 Oct. 1967, p. 6

42. Skuridin, G.A., *op. cit.*, p. 1233; *Pravda*, 22 Oct. 1967, p. 6

43. Skuridin, G.A., *op. cit.*, pp. 1236, 1250; *Pravda*, 22 Oct. 1967, p. 6

44. *Soviet Space Programs, 1971-1975*, p. 89; *Space Technology International*, Jan. 1968, p. 48

45. Doiginov, Sh., et al., "Study of the Magnetic Field from the Venera 4 Interplanetary Station"

46. Blagonravov, A.A., et al., *op. cit.*, pp. 258-259; Alekeseyev, V. and Minchin, S., *op. cit.*, p. 36; Petrov, G.I., *Conquest of Outer Space in the USSR, 1967-1970*, p. 346; *Pravda*, 22 Oct. 1967, p. 6

47. Alekeseyev, V. and Minchin, S., *op.cit.*, pp. 36. 59; Petrov, G.I., *op. cit.*, p. 347; Skuridin, G.A., *op. cit.*, pp. 1240-1242; *Spaceflight*, May 1969, p. 173; *Pravda*, 22 Oct. 1967, p. 6

48. Blagonravov, A.A., et al., *op. cit.*, p. 612; Skuridin, G.A., *op. cit.*, pp. 1228, 1237; *Pravda*, 22 Oct. 1967, p. 6

49. *Icarus*, June 1972, p. 420; Skuridin, G.A., *op. cit.*, pp. 1243-1244; *Pravda*, 22 Oct. 1967, p. 6

50. *Science*, 15 Mar. 1968; Blagonravov, A.A., et al., *op. cit.*, p. 494

51. *Spaceflight*, Jan. 1968, pp. 15-17 and Feb. 1968, pp. 49-51 and Jan. 1969, p. 14 and May 1969, pp. 171-173; Alekeseyev, V. and Minchin, S., *op. cit.*, pp. 37, 60; Petrov, G.I., *op. cit.*, pp. 338-339; Skuridin, G.A., *op. cit.*, pp. 1248-1249; Sagan, C., *The Cosmic Connection*, pp. 84-86

52. Alekeseyev, V. and Minchin, S., *op. cit.*, p. 60

53. Petrov, G.I., *op. cit.*, pp. 327-328

54. Alekeseyev, V. and Minchin, S., *op. cit.*, pp. 46,49

55. Alekeseyev, V. and Minchin, S., *op. cit.*, pp. 52-56

56. Alekeseyev, V. and Minchin, S., *op. cit.*, pp. 57-62; *Spaceflight*, Sep. 1969, pp. 326-327; *Aviation Week and Space Technology*, 23 June 1969 and 26 May 1969

57. *Aviation Week and Space Technology*, 23 June 1969; *Spacecraft*, Sep. 1969, pp. 326-327; Petrovich, G.V., *op. cit.*, p. 489; Petrov, G.I., *op. cit.*, pp. 347-349; Alekeseyev, V. and Minchin, S., *op cit.*, pp. 63-65, 71

58. Alekeseyev, V. and Minchin, S., *op. cit.*, pp. 65, 71; *Aviation Week and Space Technology*, 23 June 1969

59. Alekeseyev, V. and Minchin, S., *op. cit.*, pp. 68-72

60. Petrov, G.I., *op. cit.*, pp. 350-351; *Aviation Week and Space Technology*, 23 June 1969; Alekeseyev, V. and Minchin, S., *op. cit.*, pp. 70, 72

61. *Aviation Week and Space Technology*, 23 June 1975 and 10 Apr. 1972; *Soviet Space Programs, 1971-1975*, p. 92; *Jane's All the World's Aircraft, 1971-1972;* Petrov, G.I., *op. cit.*, p. 353; Daniloff, N., *Kremlin and the Cosmos*, p. 148

62. Ponomarev, A.N., *op. cit.*, p. 275; Alekeseyev, V. and Minchin, S., *op. cit.*, pp. 75-77

63. Petrov, G.I., *op. cit.*, p. 359; Alekeseyev, V. and Minchin, S., *op. cit.*, p. 79

64. Petrov, G.I., *op. cit.*, p. 359; Alekeseyev, V. and Minchin, S., *op. cit.*, p. 80

65. *Soviet Space Programs, 1971-1975*, pp. 91-92; Alekeseyev, V. and Minchin, S., *op. cit.*, pp. 81-83; *Icarus*, June 1972, pp. 415-461

66. Ponomarev, A.N., *op. cit.*, p. 275. *Investigations of the Upper Atmosphere and Outer Space Conducted in 1970 in the USSR (14th COSPAR)*, p. 18

67. *TRW Space Log, 1975*, p. 85; *Aviation Week and and Space Technology*, 10 Apr. 1972

68. *Jane's All the World's Aircraft, 1973-1974*, p. 652

69. *Space Research Performed in the USSR in 1972 (16th COSPAR)*, pp. 15-16; Aleseyev, V. and Minchin, S., *op. cit.*, p. 91

70. *Space Research Performed in the USSR in 1972*, pp. 18-19, 20; Alekeseyev, V. and Minchin, S., *op. cit.*, p. 93

71. *Space World*, Feb. 1973, pp. 32-33; *Space Research Performed in the USSR in 1972*, p. 17; Alekeseyev, V. and Minchin, S., *op. cit.*, pp. 85-88

72. *Space World*, Feb. 1973, pp. 34-35; *Space Research Performed in the USSR in 1972*, pp. 20-22; Alekeseyev, V. and Minchin, S., *op. cit.*, pp. 99-103; *Soviet Physics Today*, July 1973, pp. 7-9 and Nov. 1973, pp. 282-284; *Aviation Week and Space Technology*, 18 Sep. 1972, p. 23

73. *Soviet Space Programs, 1971-1975*, p. 93

74. *Spaceflight*, July-Aug. 1976, p. 299; *Aviation Week and Space Technology*, 23 June 1975, p. 20 and 22 Aug. 1977, pp. 52-54

75. *Spaceflight*, Feb. 1976, p. 4 and Apr. 1976, p. 117 and Jan. 1976, p. 40; *Science News*, 16 Apr. 1977, p. 225; *Aviation Week and Space Technology*, 26 Apr. 1976 and 17 Nov. 1975. *Soviet Space Programs, 1971-1975*, p. 144

76. *Jane's All the World's Aircraft, 1976-1977*, p. 703; *Aviation Week and Space Technology*, 28 June 1976

77. *Spaceflight*, Jan. 1976, pp. 20-21; *Aviation Week and Space Technology*, 26 Apr. 1976; *Science News*, 19 June 1976, pp. 388-389

78. *Space World*, Mar. 1976; *Spaceflight*, Jan. 1976, pp. 20-21; *Aviation Week and Space Technology*, 26 Apr. 1976

79. *Science News*, 16 Apr. 1977 and 19 June 1976; *Aviation Week and Space Technology*, 27 Oct. 1975, p. 15

80. *Science News*, 16 Apr. 1977 and 19 June 1976; *Spaceflight*, Jan. 1976, pp. 20-21; *Aviation Week and Space Technology*, Apr. 1976 and 3 Nov. 1975

81. *Jane's All the World's Aircraft, 1976-1977*, p. 703; *Spaceflight*, Jan. 1976, pp. 20-21 and July-Aug. 1976, p. 299

82. *Soviet Space Programs, 1971-1975*, pp. 142-144; *Jane's All the World's Aircraft, 1976-1977*, p. 703; *Science News*, 16 Apr. 1977; *Aviation Week and Space Technology*, 28 June 1976, pp. 60-62

83. *Aviation Week and Space Technology*, 28 June 1976, p. 60

84. *Aviation Week and Space Technology*, 27 June 1977

85. *Spaceflight*, Mar. 1979, p. 98 and May 1979, pp. 193, 210; *Aviation Week and Space Technology*, 27 Sep. 1976 and 20 June 1977 and 27 June 1977

86. *Spaceflight*, Dec. 1978, p. 402; *Aviation Week and Space Technology*, 18 Sep. 1978, p. 31

87. *Spaceflight*, May 1979, pp. 206-207 and Dec. 1978, p. 402; *Aviation Week and Space Technology*, 1 Jan. 1979, p. 21.

88. *Aviation Week and Space Technology*, 1 Jan 1979, p. 21 and 15 Jan 1979, pp. 19, 50-51.

Section IV
EXPLORATION OF MARS

INTRODUCTION

Mars has perhaps held man's interest longer than any other planet in the solar system. Its reddish color and its periods of visibility in the night sky which are longer than the brighter Venus have in part accounted for the fascination which we have always held for the planet. The observations by Lowell and the stories of H. G. Wells have fired man's imagination and guaranteed his curiosity and eventual exploration of Mars. The first attempt to send an automated probe to one of our sister planets was aimed at Mars by the USSR, a scant three years after the first artificial satellite of the earth was placed in orbit.

During the intervening 19 years, 14 confirmed Soviet spacecraft have been assigned the task of expanding our knowledge of that planet. Although the Soviet Union has not exerted as concerted an effort toward Mars as she has toward Venus, the program of Mars exploration has been a steady one employing spacecraft of ever-increasing sophistication and capabilities. Unfortunately, these automatic stations have not met with the ultimate successes of the Venera probes. Of the four attempted soft landings on Mars, all have failed. However three of the four proposed artificial Martian satellites have been placed into Martian orbits and have returned significant data.

Figure 1a summarizes these missions to the red planet. Even though the first attempt to transmit scientific information back to earth from the vicinity of Mars occurred in 1960, it was not until 1971 that any data were successfully received. The USSR did not launch spacecraft (at least which achieved earth orbit) during four of the nine launch opportunities, in contrast to missing only two of twelve launch opportunities to send probes to Venus. However, the increased reliability needed to survive the longer flight time and the greater time between launch windows can be attributed to these differences.

Figure 1b indicates the landing sites of those Soviet Mars spacecraft which have reached the surface of the planet.

MARS PROBE SUMMARY

Designator	Date Launched	Mission	Total Weight	Orbiter/ Lander Weight	Results
Unannounced	10 Oct 1960	Flyby	640 kg ?	---	Failed to achieve earth orbit
Unannounced	14 Oct 1960	Flyby	640 ?	---	Failed to achieve earth orbit
Unannounced	24 Oct 1962	Flyby	894 ?	---	Failed to leave earth orbit
Mars 1	1 Nov 1962	Flyby	893.5	---	Communications failed 21 Mar 1963 at dis- distance of 106 million km
Unannounced	4 Nov 1962	Flyby	894 ?	---	Failed to leave earth orbit
Zond 2	30 Nov 1964	Flyby	950	---	Communications failed about 4-5 May 1965
Zond 3	18 Jul 1965	Test	950	---	Photographed far side of moon; passed Mars orbit; communications failed Mar 1966
Unannounced	27 Mar 1967	Lander?	?	?	Rumored launch failure of second generation Mars probes
Kosmos 419	10 May 1971	Oribiter/ Lander	4650 ?	3440/1210 ?	Failed to leave earth orbit
Mars 2	19 May 1971	Orbiter/ Lander	4650	3440/1210	Orbiter entered 1380-2500 km orbit on 27 Nov; Lander crashed
Mars 3	28 May 1971	Orbiter/ Lander	4650	3440/1210	Orbiter entered 1500-200,000 km orbit on 2 Dec; lander survived 110 seconds on surface
Mars 4	21 Jul 1973	Orbiter	3440	3440	Engine malfunction prevented craft from entering orbit; flew by on 10 Feb 1974
Mars 5	25 Jul 1973	Orbiter	3440	3440	Entered orbit of 1760-32,560 km on 12 Feb 1974
Mars 6	5 Aug 1973	Lander	3260	1200	All signals stopped seconds before landing on 12 May 1974
Mars 7	9 Aug 1973	Lander	3260	1200	Engine malfunction caused lander to miss planet on 9 Mar 1974

Figure 1a

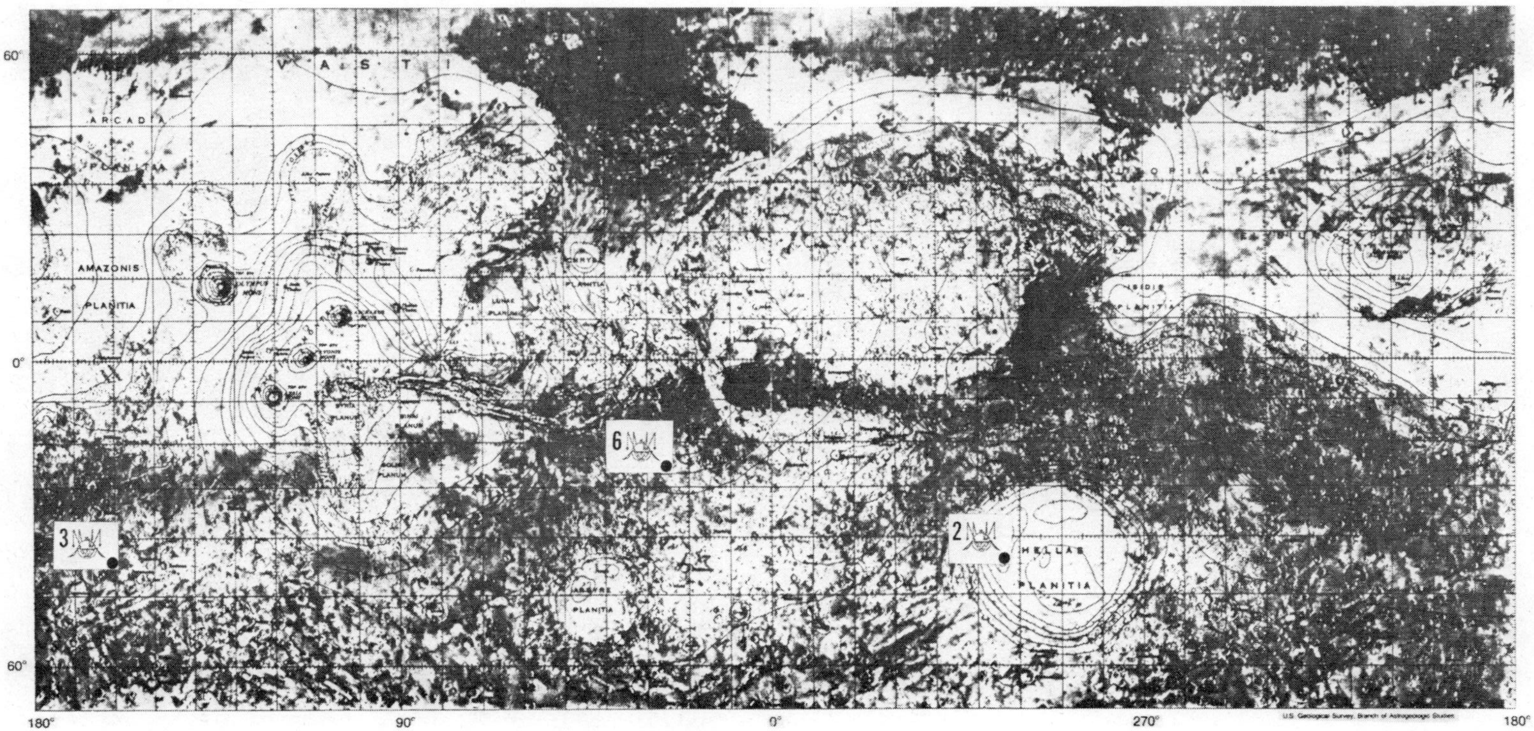

Figure 1b Location of the impacts of the three Soviet attempts to soft-land scientific capsules on Mars

Chapter 1

FIRST GENERATION SPACECRAFT

OCTOBER 1960 FLIGHTS

The first spacecraft aimed at a celestial body other than the moon were launched during October 1960 toward the planet Mars. On October 10 and then again on October 14 the Soviet Union attempted to launch the world's first interplanetary spaceships. Both probes were launched by an improved version of the standard Vostok launcher. The third stage was lengthened and was intended to act as an orbital platform from which the Mars probes were to be fired. Unfortunately both launch vehicles suffered booster malfunctions and never even attained earth orbit (this orbital platform technique was finally perfected with the Venera 1 spacecraft). These attempts have never been officially recognized by the Soviets, but were publicly disclosed by US authorities in September 1962.[1]

An interesting historical event was also taking place at this time. Nikita Khruschev was enroute by ship to address the United Nations in New York. After his ship had docked, a disgruntled Soviet seaman managed to defect to US authorities. Many rumors regarding these October space launchings have been attributed to this seaman. Everything from earth orbital bombs to giant lunar rockets to manned spaceships have been reported to have been attempted in order that Khruschev might take advantage of his audience at the United Nations and before the world. The seaman is reported to have seen replicas of an "entirely new type of spaceship" to be unveiled if the flights were successful. It seems now that if there is any merit to these stories, they probably revolve around the first flybys of Mars.

It is extremely difficult to speculate on the configuration of these two Mars probes since the Soviets have refused to admit their existence let alone release details as to their construction. Four months later saw the launching of the 643.5-kg Venera 1. As has been seen, this craft differed markedly from the succeeding Venus probes of Zond 1 and Venus 2-3. However, the first admitted Mars probes (Mars 1, Zond 2-3, to be discussed in detail later) closely resembled these Venus probes. This fits in well with the Soviet desire to standardize whenever possible. Therefore without evidence to the contrary one may reasonably assume that the October 10 and 14 flights carried probes quite similar to Venera 1 and not until 1962 were the advanced 900-950-kg probes introduced.

MARS 1 FLIGHTS

Since the sidereal period of Mars (the time Mars takes to revolve once around the sun with respect to the stars) is approximately 687 days whereas the sidereal period of Venus is only 225 days, opportunities to launch spacecraft to the red planet occur less frequently than to Venus--approximately every 25 months for Mars versus 19 months for Venus. Hence the next launch window following the October 1960 attempts came in October-November of 1962.

On 24 October 1962 an unannounced satellite was placed into an earth orbit of 485 km by 180 km with an orbital inclination of 64.9°.[2] The evidence points to this being the first attempt to launch a spacecraft to Mars during the 1962 window. The probe, however, failed to leave its orbital platform, but appeared to explode and decayed five days later. As has already been demonstrated with the Venus probes and the 1960 Mars attempts, the Soviets followed the planetary multi-probe concept. Therefore, on 1 November a second satellite was placed into an earth orbit of 243 km by 174 km with 64.9° inclination. This time, though, the spacecraft was successfully fired from orbit into a Martian trajectory and was designated Mars 1. Three days later, on 4 November, a final attempt was made to complete a twin set of probes to Mars. This shot failed, too, when the intended "Mars 2" spacecraft did not leave earth orbit. It was destroyed while reentering the earth's atmosphere the next day.[3]

Figure 2 illustrates the virtually identical design of Mars 1 compared with Venera 2 and 3 and Zond 1-3, and Figure 3 shows the trajectory of Mars 1 from earth to Mars. The spacecraft measured 3.3 m in length, 1.1 m across the orbital compartment and 4 m across the extended solar arrays and hemispherical thermal radiators. It weighed 893.5 kg. The 2-m diameter parabolic communications antenna was designed for operation

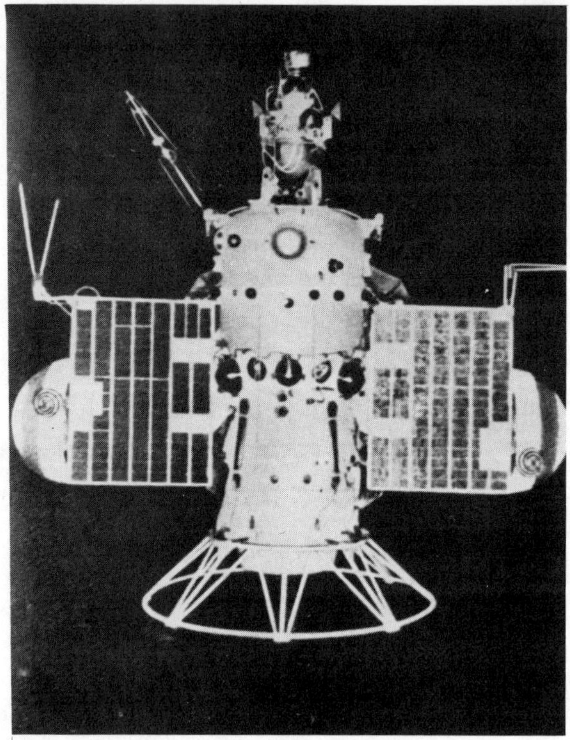

Figure 2a Photograph of the Mars 1 spacecraft launched on 1 November 1962

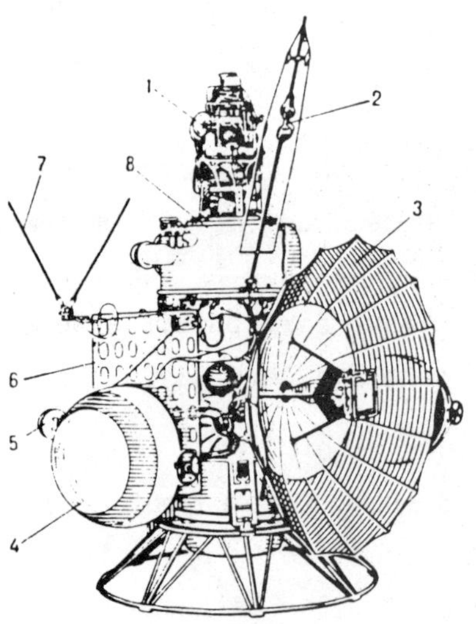

1—vernier rocket engine; *2*—magnetometer probe; *3*—high-gain antenna; *4*—temperature control system radiator; *5*—semi-directional antenna; *6*—solar panel; *7*—omni-directional antenna; *8*—orbital compartment.

Figure 2b Schematic diagram of the Mars 1 interplanetary station

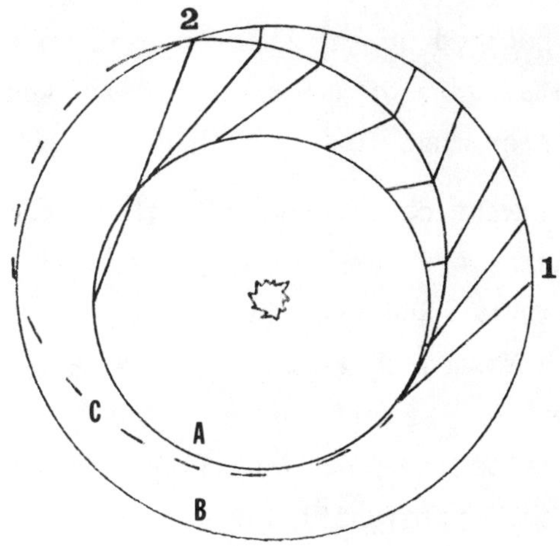

A. Orbit of the earth
B. Orbit of Mars
C. Orbit of the Mars 1 spacecraft

1. 1 November 1962, Position of earth and Mars at the launch of Mars 1
2. 19 June 1963, Position of earth and Mars at Mars 1 encounter

Figure 3 Trajectory of Mars 1 from the earth to Mars

in the vicinity of Mars where weak signal strengths require highly directional antennae. Pressure within the orbital compartment housing and most of the scientific apparatus was maintained at 850 mm Hg while the temperature varied from 20° to 30°C.[4] The temperature control system relied on the two hemispherical attachments at the end of each solar panel. As with the Venera 2 and 3 spacecraft each radiator was composed of two sections which had different heat-absorption characteristics. By transferring heat via a fluid from the orbital compartment to a section of the radiator or vice versa, the temperature within the compartment was maintained.[5]

The array of scientific instruments was varied. Carried on board were magnetometers for detecting Martian and interplanetary magnetic fields, scintillation and gas-discharge counters to detect any possible Martian radiation belts and cosmic ray spectrums, a spectrograph for examining absorption bands of ozone in the planet's atmosphere, a spectro-reflexometer for discovering possible organic compounds on the surface of Mars, charged particle traps for analyzing cosmic rays, and a radio telescope operating at 150 and 1500 m wavelengths to record cosmic radio-emissions.

In addition, a photo-television system was available to transmit the first close-up pictures of the Martian surface. Photographs taken by this system were to be developed on board the spacecraft and then scanned by a TV camera for transmission to earth. Located on the solar panels on the side opposite the solar cells was a total area of 1.5 m^2 which was sensitive to micrometeors with masses greater than 10^{-7} g .[6]

Communications between earth and Mars 1 were carried out via three transmitter-receivers, operating at wavelengths of 5 and 8 cm, 32 cm, and 1.6 m. Sessions were held at intervals of first two and then later five days. Unfortunately all contact was lost with Mars 1 after almost five months of flight on 21 March 1963 at a distance of 106 million km. The loss of signals was apparently due to a malfunction in the orientation system of the spacecraft.[7] However, a great deal of significant data was received prior to this date and the flight was by no means a complete failure.[8] Mars 1 is estimated to have passed the red planet on 19 June 1963 at a distance of about 193,000 km.[9]

ZOND 2

The next Soviet Mars probe came with the launching of Zond 2 on 30 November 1964, probably to attempt a close photographic flyby the day after the US Mariner 4 was to reach the planet.[10] The 950-kg spacecraft closely resembled Mars 1. Unexpectedly on 2 December Pravda announced that the power output on board the spacecraft was only one-half the anticipated level. This power reduction was probably caused by the failure of one solar array to deploy or function properly. Even with this handicap, contact was maintained with the craft until about 4-5 May 1965. Zond 2 is reported to have missed Mars by 1500 km on 6 August 1965.[11]

Of particular note concerning Zond 2 is the first use in space of electric plasma engines for attitude-control purposes. Six such engines (in addition to the usual attitude system) were arranged to control the ship's orientation to determine the feasibility of plasma jets on longer duration flights.[12] The term "plasma engine" as used by the Soviets refers to either electromagnetic or electrostatic ion engines. In the case of Zond 2 the engines were of the electromagnetic type. A working fluid is first converted into a plasma and then accelerated to velocities of up to hundred of km/sec via an electromagnetic field. Plasma engines of the electrostatic variety were first used on the Voskhod manned spaceships.[13] Very few results from the scientific instruments aboard Zond 2 have been published. Due to the reduced power capabilities, most experiments were probably drastically curtailed with the hope of conserving power until the encounter with Mars.

ZOND 3

Zond 1 had been sent to investigate Venus whereas Zond 2 had been directed to Mars. The next launch window for spacecraft approaching Mars would not occur until January 1967 and the next opportunity for sending a probe to the vicinity of Venus would not arrive until November 1965. Thus, it came with great surprise when on 18 July 1965 the Soviet Union announced the launching of Zond 3 for the purpose of "testing of the station's systems in actual conditions of a prolonged space flight and scientific research in the interplanetary space."[14] Since a rendezvous with either Mars or Venus at this time was impossible, what could the real purpose of the flight have been?

The answer to this question came very quickly. Thirty-three hours after launching, Zond 3 passed by the far side of the moon, approaching as close as 9220 km, and in one hour and eight minutes photographed a substantial portion of the lunar far side. Zond 3 then proceeded on a heliocentric path away from the sun and earth.[15] The purpose of the mission was now apparent: to test the spacecraft systems which had previously failed on Venera 1, Zond 1-2, and Mars 1 before the next Venus and Mars launch windows. Zond 3 may have been the companion to Zond 2, whose launch was scrubbed due to the power problems experienced by Zond 2. That would also explain why Zond 2 has remained the only single probe attempt in the entire Soviet Venus and Mars programs.

Zond 3 was virtually identical with the Mars 1 and previous Zond probes (and with the Venera 2 and 3 probes to follow). A photograph of Zond 3 is provided in the Zond section of this book. It carried the same array of instruments as detailed during the descriptions of the Mars 1 and Venera 2-3 stations, including the plasma engines of Zond 2.[16] However, on this occasion the photo-television system was finally given the opportunity to prove itself. For a more detailed description of this system and its significant results the reader is referred to the lunar exploration (Zond series) section of this book.

The relatively high quality photographs were transmitted from Zond 3 on 29 July at a distance of 2.2 million km from the earth and again at a distance of 31.5 million km.[17] Hence the reliability of the photo-television system was established at a distance roughly equivalent to that needed by the Venera 2 spacecraft which was to be launched the following November. Zond 3 continued along its heliocentric orbit and approached the orbit of Mars, which, of course, was not in the vicinity since the launch did not occur at the proper time. Contact was maintained with the station until March 1966 when the probe was 150 million km from the earth.[18]

Considering the partially successful flights of Veneras 2 and 3 in 1956-1966 (both probes functioned properly until just before reaching Venus) and, in restrospect, the fact that Venera 4 and Kosmos 167 were launched toward Venus in June 1967, it is puzzling that when the next Mars launch window arrived in January 1967 no spacecraft were detected placed in orbit that may have been failed Mars-type probes. In addition, no substantial rumors emerged referring to launch failures of these probes. One report places financial difficulties as the cause of this hiatus of Martian exploration.[19] For whatever reasons, one must assume that the Soviets had decided not to attempt to send a probe to Mars, marking the first Venus or Mars planetary window not utilized since the very first interplanetary probes were launched in October 1960.

After accepting the fact that the USSR had chosen to skip the 1967 opportunity, Western observers started looking toward the Febraury-March 1969 window with expectations of seeing a new, more advanced, type probe. They were assured of the Soviets' determination to explore outer space with the flights of Venera 4 (June 1967), Zond 4-6 (1968), and the recent flights of Veneras 5 and 6 (January 1969), and the manned Soyuz 4-5 mission (January 1969). However, February and March came and passed without a single Kosmos flight which could possibly be attributed to an unsuccessful Mars attempt. Eleven Kosmos satellites were orbited during this period but all could reasonably be assigned other missions. This time, though, rumors did arise that a new, more powerful rocket had attempted to launch two Mars spacecraft, but both experienced booster failure and never achieved earth orbit. The first attempt was reported to have occurred in late February or early March with the second attempt taking place on 27 March.[20] The basis behind these rumors was the assumption that the new heavy Proton-class launcher would be employed to send a massive, more advanced spacecraft to possibly soft-land on Mars. This new launch vehicle made its debut on 16 July 1965 by placing the 12,200-kg Proton 1 satellite into earth orbit. The Zond 4-6 lunar probes of 1968 demonstrated the capabilities and apparent reliability of the Proton rocket as a carrier of large scientific probes. The speculations surrounding these possible attempts to investigate Mars in 1969 remain unconfirmed.

Chapter 2

SECOND-GENERATION SPACECRAFT

KOSMOS 419, MARS 2, MARS 3

Finally, during the 1971 opportunity, the long-awaited second generation of Soviet Martian space probes appeared on the scene, not however without an anxious beginning. On 10 May 1971 a Proton launch vehicle placed a heavy payload into an earth orbit of 174 km by 159 km with an inclination of 51.4°. Unfortunately, the spacecraft which would have been Mars 2 did not restart and remained with its launch platform. The spacecraft was subsequently designated Kosmos 419 and was destroyed while reentering the atmosphere two days later.[21]

Nine days after the Kosmos 419 launch, a Mars 2 spacecraft was rocketed into an earth orbit of 173 km by 137 km at an angle of 51.8° and then redirected on a trajectory to the planet Mars. After a lapse of six and one-half years, the Soviet Union had a probe on the way to the red planet. To insure a successful mission during this launch window, a Mars 3 station followed Mars 2 on 28 May 1971. The Soviet pattern of dual planetary probes was thus complete.

The new Mars spacecraft represented a major new design in interplanetary probes (Figure 4). They were to become the forerunners of a new series of stations which included Mars 4-7 and Venera 9-12. Each craft weighed approximately 4650 kg--almost five times the weight of the last true Mars probe, Zond 2. Each craft also had a dual mission: soft-land a descent capsule on the surface of the red planet and place into orbit a photographic and scientific satellite.

As with all previous probes of an interplanetary nature, Mars 2 and 3 were divided into two major modules. The first of these was the orbiter/bus which functioned to deliver the landing payload to the vicinity of Mars and which then would become an artificial satellite of the planet. The 4650-kg weight of the spacecraft is thought to be broken into approximately

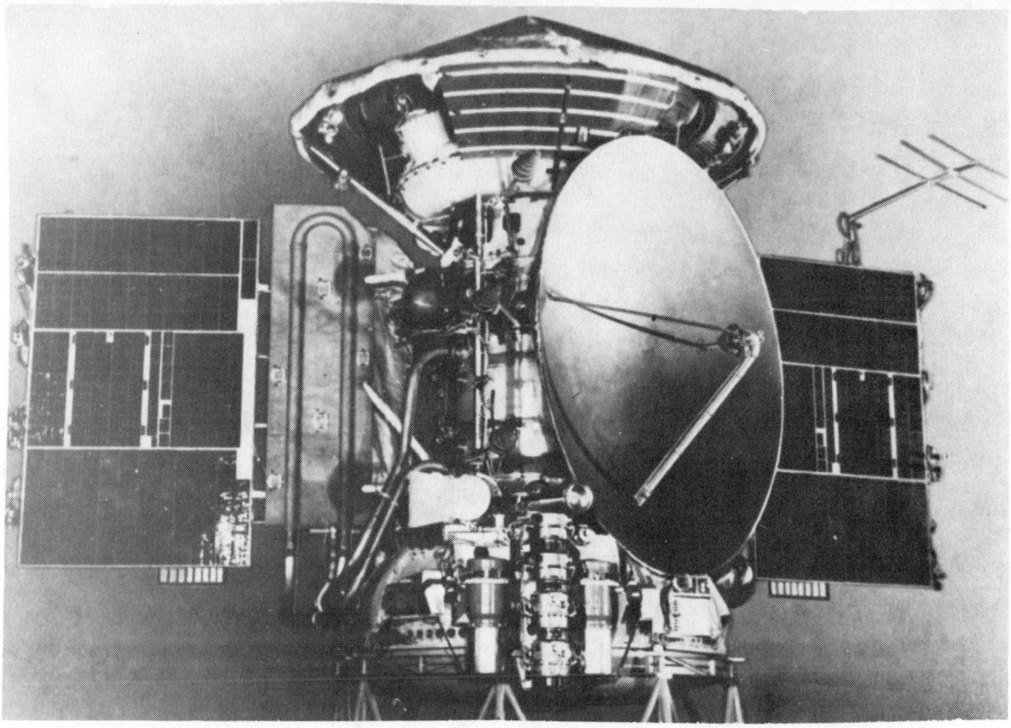

Figure 4a Configuration of the Mars 2/3 second generation Mars explorers (TASS from SOVFOTO)

3440 kg for the fully fueled orbiter at launch from earth, and 1210 kg for the entire descent apparatus, 350 kg which consisted of the actual landing capsule. Each craft stood approximately 4.1 meters high with a base diameter of 2 m and a breadth of 5.9 m across the solar panels. A 2.5 m diameter parabolic antenna provided high-gain communications.[22]

The orbiters contained the following scientific experiments:

1. Apparatus to measure fluxes and energies of solar wind particles
2. A three-component magnetometer to determine interplanetary and Martian magnetic fields
3. A French-built Stereo 1 experiment (Mars 3 only) designed to monitor solar radiation in metric wavelengths simultaneously with earth-based receivers to determine the mechanisms of solar outbursts
4. A 1-kg infrared radiometer to measure radiation emitted in the 8-40 micron range to determine the temperature of the Martian surface down to -100°C
5. A radiotelescope to record polarization and intensity of 3.4-cm emissions (to measure the temperature and dielectric constant of the soil at depths of 3-50 cm below the surface of the planet)

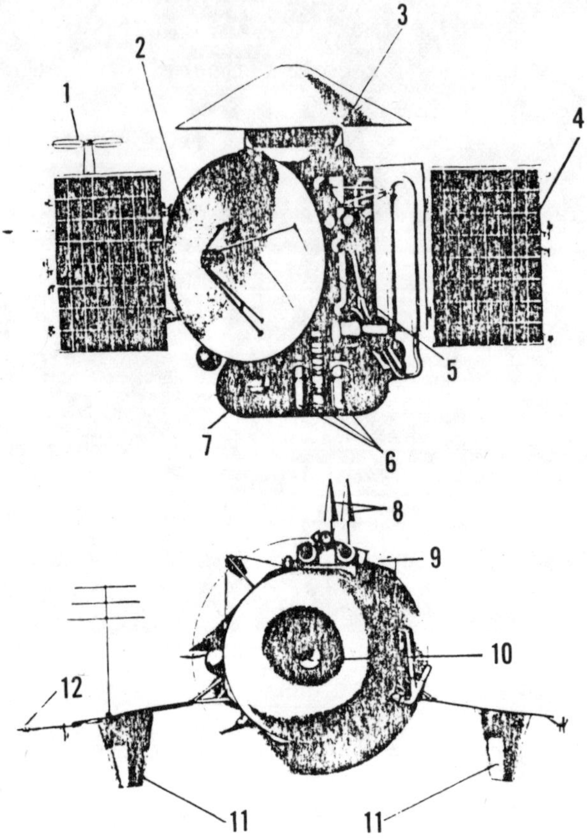

1. Antenna of scientific apparatus "Stereo" (Mars 3 only)
2. Parabolic communications antenna
3. Descent apparatus
4. Solar panel
5. Main body of spacecraft (propulsion engine tank)
6. Optical electronic instruments of the astro-orientation system
7. Instrument compartment
8. Low-power directional antennae
9. Optical electronic instruments of the navigation system
10. Main correction and braking engine
11. Antennae for communications link with the descent apparatus
12. Magnetometer

Figure 4b Diagram of the Mars 2/3 spacecraft

6. CO_2 gas absorption strips (2.06 microns) determined the relief of the surface from the optical thickness of the atmosphere.
7. Ultraviolet photometers recorded amounts of atomic hydrogen, oxygen, and argon.
8. A 1.38-micron wavelength photometer investigated presence of water vapor in the atmosphere.
9. The structure of the atmosphere was detailed by refraction of the crafts' transmissions when eclipsed by the planet.
10. Photoelectric measurements of the brightness of the surface were registered in six narrow ranges between 3700 Å and 7000 Å.
11. Each craft contained a pair of cameras for photographing the clouds and surface of Mars (each camera pair contained both a wide-angle and a telephoto lens).[23]

The majority of this equipment was located at one end of the orbiter in or on the inverted top-like compartment through which the nozzle of the correction engine extended. Above this compartment was the cylindrical body of the station containing the liquid fuel for course corrections and orbital insertion. Two solar panels, the high-gain parabolic antenna, and other smaller antennae were attached to this body, while the two large antennae extending beneath the solar panels were for communications between the lander and orbiter.

The descent apparatus was fastened to the end opposite the propulsion engine nozzle and consisted of a landing capsule, a parachute system, a 2.9 m diameter aerodynamic brake shield, and a retro-rocket system. The actual landing sequence will be described in detail later. The 1.2-m landing capsule carried instruments similar to those carried on the Venera probes and included sensors to measure the temperature, pressure, and wind velocities of Mars. In addition, a mass spectrometer could determine atmospheric composition. Other instruments were to measure the mechanical and chemical properties of the soil, while major importance was attached to the inclusion in the landing capsule of two television cameras capable of a 360-degree scan of the Martian surface.[24] Thus, the Martian landing capsule was remarkably similar to the Luna landing capsules which had touched down on the lunar surface five years before (Figure 5).

The Mars 2 spacecraft continued toward its rendezvous with Mars, making course corrections on 17 June, 20 November, and 27 November.

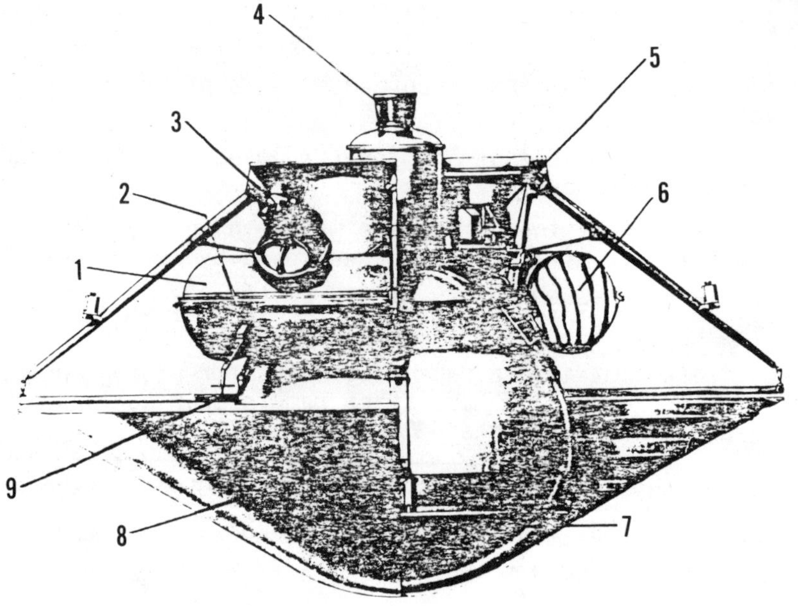

1. Parachute container
2. Antenna for communications link with the Mars orbiter
3. Braking parachute deployment engine
4. Descent apparatus separation engine
5. Automatic control instrumentation
6. Main parachute
7. Landing capsule
8. Aerodynamic braking shield
9. Radioaltimeter antenna

Figure 5 Cutaway of the Mars landing system

During this flight, duplicate Mars 2 and Mars 3 probes were operated in an environmental chamber on earth as had been done with the Venera probes to aid in correcting any problems which might arise. Again following the procedures of the Venera probes, the batteries within the landing capsule were charged only shortly before the separation from the mother craft. On 27 November 1971 the descent capsule of Mars 2 was jettisoned 4.5 hours before reaching the planet. The orbiter continued along its path, completed an orbital-insertion burn and entered a 1380 km by 25,000 km orbit about Mars at an inclination of 48° 54' to the equator, resulting in an orbital period of about eighteen hours. Unfortunately, after entering the Martian atmosphere at 44.2°S and 313.4°W the descent module apparently

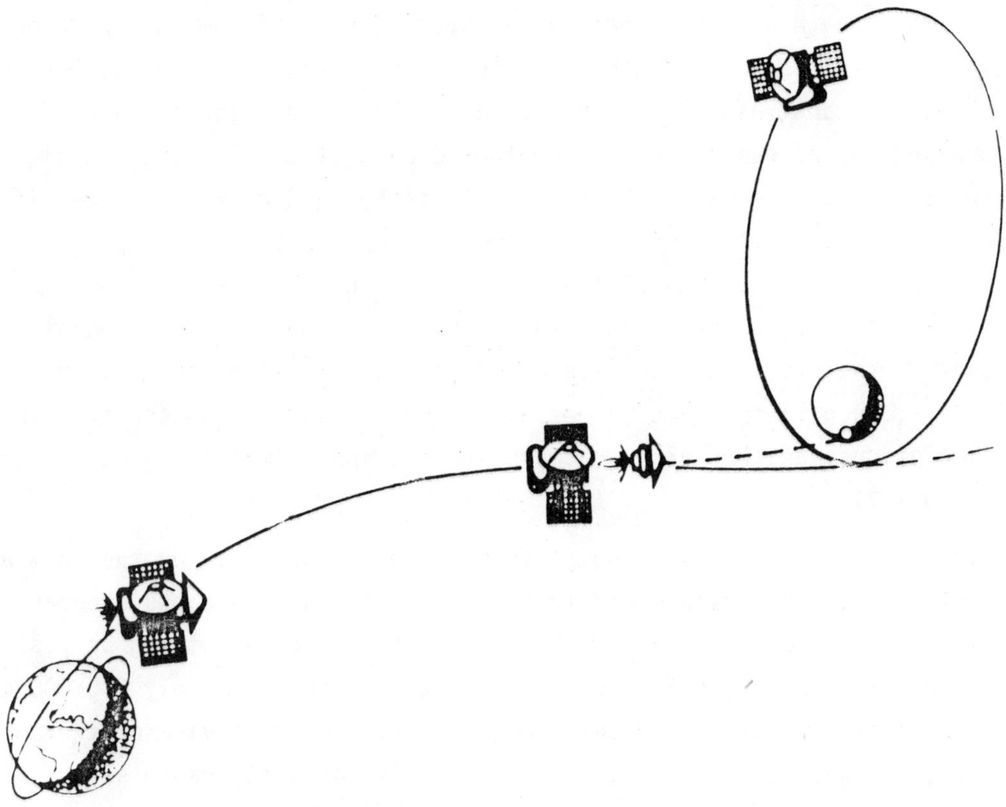

Figure 6 Trajectory of Mars 2/3 from the earth to Mars

suffered a malfunction during the landing sequence and was destroyed upon impact with the Martian surface near 45° S, 302° W (Figure 6).[25]

Although the Mars 3 spacecraft had been launched nine days after Mars 2, it arrived at the red planet only five days after its predecessor, following a course correction on 8 June. The descent module was detached at 1214 hr Moscow time on 2 December 1971, and fifteen minutes later the module fired its solid fuel descent engine and aligned its axis to present its aerodynamic braking shield forward. The module entered the Martian atmosphere at 1644 hr with a velocity of 6 km/sec at an angle of less than 10 degrees to increase drag. An overload sensor activated the braking parachute while the craft was still traveling at supersonic velocity through the rarefied atmosphere. The main chute followed but was reefed until the velocity of the craft had slowed to sonic velocity at which time the main chute was fully deployed. Upon opening of the main parachute, the heat shield was discarded and the radio altimeter antennae were deployed and activated. At a height of 20-30 m above the surface, a final braking rocket was fired to slow the descent of the landing capsule. At the same time, the main parachute was detached and carried to one side by a smaller rocket to prevent its landing atop the capsule. Special shock absorbers within the capsule helped to lessen the impact further, which is said to have occurred at a velocity of 20.7 m/sec. Upon reaching the surface the sterilized capsule unfurled its petal-like cover to right itself, much in the same manner as the Luna 9 and 13 spacecraft. Total time of descent through the atmosphere was approximately three minutes (Figure 7).[26]

Ninety seconds after landing (the location of the probe on the Martian surface was later determined to be 45° S, 160° W) the capsule began transmitting data to earth via the Mars 3 orbiter which had taken up position around Mars in an orbit between 15,000 km and approximately 200,000 km. Incorporated within these data was the first panoramic view of the surface. Fate, though, struck again. After twenty seconds of transmissions all signals ceased. For some inexplicable reason the Mars 3 capsule was never heard from again. The only partial panoramic photograph disappointedly revealed no contrasts and indicated an illumination of only 50 lux.[27]

Many explanations have been given for the failure of the Mars 3 lander. One of the most prevalent hypotheses has been that the lander was blown over. Unluckily, Mars 2 and 3 arrived at the red planet during one of the worst dust storms in recorded history. The entire planet was enveloped in the swirling clouds of dust, obliterating any recognizable features on the surface. Mars 2 and 3 did not have the capability of entering orbit and ejecting the landing module at a later, more convenient time (as did the US Viking spacecraft four and one-half years later). The Soviets had no choice--the landers had to take their chances in the storm. It is quite possible that the dust storms were also responsible for the failure of the Mars 2 capsule.[28]

An alternate explanation has been put forth by two Soviet space scientists M. Ya. Marov and G. I. Petrov. According to these highly respected spokesman, the Mars 3 orbiter itself was the culprit when it ceased

A. Mars orbiter after separation of descent apparatus
B. Descent apparatus pulling away from Mars orbiter
C. Firing of main descent engine for entry into the Martian atmosphere
D. Deployment of drogue parachute
E. Separation of drogue parachute and upper portion of main parachute housing
F. Deployment of main parachute
G. Detachment of main parachute and ignition of final braking rocket
H. Separation of landing capsule
I. Landing capsule on the Martian surface with deployed instrumentation

Figure 7 Landing sequence of second-generation Mars spacecraft (R. Escarcega)

relaying transmissions from the Mars 3 lander to earth. Thus, the descent capsule may indeed have performed history's first soft landing on Mars, but could no longer tell us of its achievements. Whatever the reason for the silence from the surface of Mars, what had begun as one of the most ambitious efforts of planetary exploration ended in great disappointment.

The missions of Mars 2 and 3, however, were far from failures. The two orbiters continued to send back data for several months. The 18-hour period of Mars 2 and the 12-day period of Mars 3 enabled a wide variety of measurements to be made. As a rule the scientific instruments were activated for only 30 minutes near periapsis. Surface temperatures ranged from +13° C at midday to -110° C at night and pressures varied from 5.5 to 6 millibars. In August 1972 the Soviets announced that Mars 2 and 3 had completed their missions.[29]

MARS 4, MARS 5, MARS 6, MARS 7

The next Martian launch window occurred in July-August 1973. During this period the world witnessed an unprecedented effort to investigate the planet named for the god of war. Like shells from a cannon the Soviet Union launched one spacecraft after another. No less than four probes were placed into earth orbit and subsequently were fired to a rendezvous with Mars seven months later. These automatic stations were designated Mars 4, 5, 6, and 7 and were launched by Proton boosters on 21 July, 25 July, 5 August, and 9 August, respectively.

The reason for sending such a large number of probes at one time can be found in the mechanics of oppositions between Mars and earth. The great opposition of 1971 allowed the Proton launch vehicle to place a large scientific payload in the vicinity of Mars. However, in 1973 more fuel was required to reach the fourth planet from the sun. Since the launch vehicle was already working at capacity, the extra weight of fuel was added at the expense of the scientific payload. It was now impossible to combine both an orbiter and a lander in one space vehicle. Hence the mission of Mars 4 and 5 was to place new artificial satellites in orbit around Mars. Mars 6 and 7, on the other hand, were to soft-land capsules on the Martian surface while conducting a flyby of the planet about one month after the arrival of Mars 4 and 5.[30]

The same basic spacecraft developed for Mars 2 and 3 and Veneras 9 - 12 were used for Mars 4 - 7. Mars 6 and 7 were virtually identical in appearance to Mars 2 and 3, whereas Mars 4 and 5 were similar sans the descent module. Although little mention has been made by the Soviets of the weights of Mars 4 - 7, it appears that Mars 4 and 5 were almost duplicates of the Mars 2 and 3 orbiters. Their fully fueled weight has been estimated to be 3440 kg with about 2270 kg dedicated to the weight of the probes in orbit. Likewise, Mars 6 and 7 each weighed approximately 3260 kg of which 1200 kg comprised the landing module (Figure 8).[31]

Mars 4 and 5 carried instrumentation very similar to the orbiters of Mars 2 and 3. These included:

1. Instrumentation for radioprobing of the atmosphere at wavelengths of 8 and 32 cm
2. A radiotelescope operating at 3.5 cm
3. An infrared radiometer operating in the range 8 - 26 μm
4. A spectrophotometer with interference filters operating in the range 2 - 5 μm
5. A narrow-band photometer with interference filters for CO_2 band near 2 μm
6. A narrow-band interference-polarization photometer for the H_2O band 1.38 μm
7. A photo-televison complex
8. A photometer with interference filters in the range 0.3 - 0.8 μm
9. Two polarimeters making it possible to measure the degree of polarization in nine narrow bands from 0.35 to 0.8 μm
10. A photometer for the ozone band 2600 Å
11. A photometer for measuring the intensity of scattered solar radiation in the Lyman-alpha line λ1215 Å
12. A gamma spectrometer for measuring gamma radiation of the planet and the cosmic background along the flight trajectory.[32]

The above-listed equipment was to be used to investigate the planet itself. In addition, Mars 4 and 5 carried a magnetometer, plasma traps and a multichannel electrostatic analyzer to record data during the flight to Mars and in the space surrounding the planet.[33]

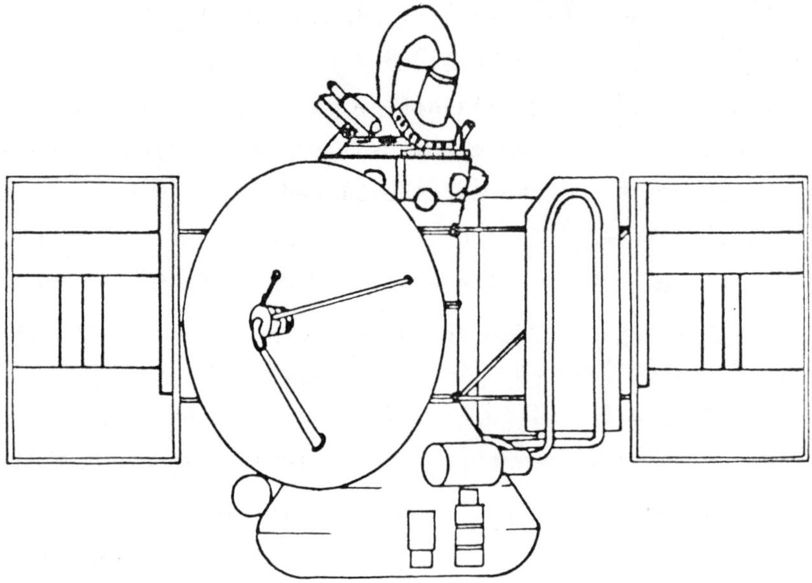

MARS 4/5

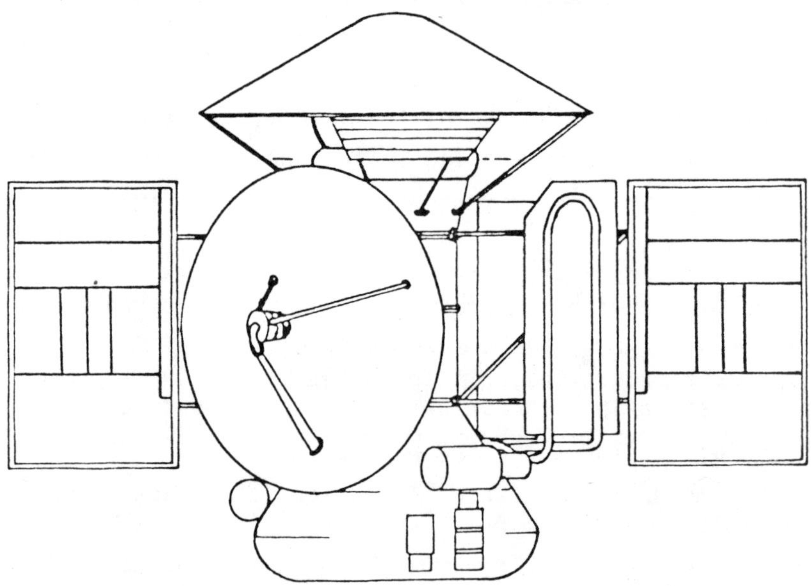

MARS 6/7

Figure 8 Comparison of the Mars 4/5 and Mars 6/7 spacecraft sent to Mars in 1973 (R.F. Gibbons)

The Mars 6 and 7 landing modules were similar to their counterparts on Mars 2 and 3, but had been increased to 635 kg. Sensors for measuring temperature, pressure, and accelerations, a radioaltimeter, and a mass spectrometer to evaluate atmospheric composition were carried. Instruments to analyze the Martian soil were to be placed on the surface after landing.[34] One Soviet scientist reported that one of the craft was to land near the Martian south pole and would relay pictures of the surface.[35]

The Mars 6 and 7 buses were scheduled to perform a number of experiments during the journey to Mars and as they bypassed the planet at close range. On board were magnetometers and plasma traps similar to Mars 4 and 5. In addition, they held micrometeorite and cosmic-ray sensors. Mars 7 carried French equipment similar to the Stereo apparatus on Mars 3. It has also been reported that both Mars 6 and 7 contained a French experiment called Zhemo whose purpose was to study solar proton and electron fluxes.[36]

All four spacecraft carried a two-circuit gas circulation system as part of their thermal control system. The gas within the orbital compartment was circulated by fans. Working in conjunction with two heat exchangers, this gas was able to maintain the interior of the compartment between approximately 20°C and 5°C. During the flight from earth to Mars the decrease of solar flux striking the solar panels resulted in a temperature drop from approximately 90°C near earth to 20°C near Mars. However, by rearranging the manner in which groups of solar cells were connected the voltage and current of the spacecraft systems could be regulated.[37]

Mars 4 was the first to reach the vicinity of the red planet after a course correction on 30 July. As Mars 4 approached the planet on 10 February 1974, the retro-engine intended to place it into orbit around Mars malfunctioned and the craft continued along its heliocentric trajectory, passing the planet at a distance of 2200 km. The spacecraft was, however, able to photograph Mars as it passed, and to relay these photos to earth.[38]

The Mars 5 station proved to be more successful when it arrived on 12 February after a course correction on 3 August. Its engine correctly fired at 1845 hr Moscow time and the craft entered an orbit about Mars of 1760 km by 32,560 km with an inclination of 35°20' and orbital period of 24 hours 52 minutes 50 seconds. Although Mars 4 was to have been the radio link between the Mars 6 lander and earth, Mars 5 was positioned to pass over both the predicted sites of Mars 6 and 7 and could complement both landers.[39]

Mars 7 which was to work in conjunction with the Mars 5 orbiter arrived ahead of Mars 6 following a course correction on 16 August. Its landing module was released by the mother craft a few hours before its scheduled entry into the atmosphere. To the great dismay of Soviet scientists, the landing module missed the planet entirely and flew by at a distance of 1300 km on 9 March. The Mars 7 bus like Mars 6 was equipped with instruments to examine Mars in a flyby mode and this task was completed.[40]

After a course correction on 13 August, the fourth and final probe, Mars 6, encountered Mars on 12 March 1974. Its batteries having been charged a month earlier, the descent module was separated at 0802 hr. Just over four hours later at 1206 hr the probe entered the Martian atmosphere with a velocity of 5.6 km/sec. Its descent sequence was virtually identical with that of the Mars 3 lander and at 1211 hr the capsule impacted on Mars at 24° S, 25° W. (The Soviets had previously requested all data from the US Mariner 9 spacecraft concerning this area and another at 43° S, 42° W.). As if the planet was determined to withhold its secrets, all signals from the capsule ceased seconds before touchdown, 148 seconds after the opening of the main parachute.[41]

From a precursory analysis, the Mars 4 - 7 probes appear to be a highly expensive disaster. Two of the four probes passed the planet without performing their missions. Communications with a third were broken moments before landing was to have occurred. Only Mars 5 which went into orbit around Mars seemed to have successfully carried out its task. However, a quantity of interesting data was returned. The Mars 6 and 7 buses did complete their bypass experiments while Mars 4 also was able to examine and photograph the planet during its unexpected path by the planet. Even the Mars 6 lander was able to take readings during its descent through the atmosphere. Among many other results, the average temperature and pressure of the Martian surface was determined to be -43° C and 6 millibars, respectively.[42]

August of 1975 presented the next launch window for spacecraft to explore Mars. The much publicized US Viking lander/orbiter probes were launched during this period and Western observers had anticipated a renewed effort by the Soviet Union at the same time. Much serious talk by Soviet scientists had been given to Mars rovers, soil-sample return probes, and even combination rover - return craft.[43] The launch window came and passed, though, without any apparent Soviet attempts. Explanations ranging from budgetary problems to political motives (for instance, spacecraft might fail or seem inferior to the US Viking) to a research delay to perfect more ambitious explorers have been given.

Likewise the October-November launch window of 1977 was not utilized for any further Mars probes. Again redesign efforts were usually proposed for the absence of spacecraft. The similarity between these two missed opportunities and the brief hiatus during the 1967 and 1969 Martian opportunities did not go unnoticed.[44]

Now, with yet another opportunity about to present itself, indicators are that the late 1979 Mars launch window will be put to good use by the Soviets. However, without an improved D-class rocket one would not imagine a significant increase in lander sophistication. While Mars rovers and sample soil return probes may be high on the list of Soviet space scientists, such robot explorers appear beyond the capacity of the present interplanetary launch vehicle.

Regardless of the reasons, one may rest assured that the recent hiatus in Martian exploration is only a temporary one. The Soviets have made it known that they are far from finished with our red planetary neighbor--probably the first planet which will bear the footprints of man other than our native earth.

REFERENCES AND NOTES

1. *Space World,* Jan. 1975, p. 9; *Aviation Week and Space Technology,* 17 Sep. 1962, p. 21; *Soviet Space Programs, 1971-1975*, p. 85; *Science,* 25 Feb. 1966; *Spaceflight,* Mar. 1977, p. 118; Turnill, R., *The Observer's Book of Unmanned Spacecraft,* pp. 156-157

2. *Soviet Space Programs, 1971-1975,* p. 556

3. *TRW Space Log, 1975,* p. 44; *Science,* 25 Feb. 1966; *Soviet Space Programs, 1971-1975,* pp. 86, 556; Stoiko, M, *Soviet Rocketry: Past, Present, and Future,* pp. 147-418; Gatland, K., *Robot Explorers,* pp. 169-170

4. *Jane's All the World's Aircraft, 1963-1964; Science,* 25 Feb. 1966. Petrovich, G.V., *The Soviet Encyclopedia of Space Flight,* pp. 249-250; Skuridin, G.A., *Mastery of Outer Space in the USSR, 1957-1967,* pp. 1143-1144

5. Gatland, K., *op. cit.,* p. 172

6. Skuridin, G.A., *op. cit.,* p. 1145; *Jane's All the World's Aircraft, 1963-1964;* Wukelic, G.E., *Handbook of Soviet Space-Science Research,* pp. 83, 88

7. Skuridin, G.A., *op. cit.,* pp. 1147, 1149; *Soviet Space Programs, 1971-1975,* p. 87; *Jane's All the World's Aircraft, 1963-1964;* Petrovich, G.V., *op. cit.,* p. 250; *Science,* 25 Feb. 1966; Blagonravov, A.A., et al., *USSR Achievements in Space Research,* p. 613; Shelton, W., *Soviet Space Exploration: The First Decade,* pp. 219-220

8. Wukelic, G.E., *op. cit.,* pp. 170, 261, 278, 286; Skuridin, G.A., *op. cit.,* pp. 1149-1150; Blagonravov, A.A., et al., *op. cit.,* p. 613

9. *Soviet Space Programs, 1971-1975,* p. 87

10. Possible impact attempt. *Science,* 24 Mar. 1967

11. Skuridin, G.A., *op. cit.,* p. 1152; *New York Times,* 2 Dec. 1964; *Soviet Space Programs, 1971-1975,* p. 87; Wukelic, G.E., *op. cit.,* p. 74; Turnill, R., *op. cit.,* p. 184; Gatland, K., *op. cit.,* p. 184; Gatland, K., *op. cit.,* p. 177

12. Petrovich, G.V., *op. cit.,* p. 509; *Space World,* Mar. 1976; Skuridin, G.A., *op. cit.,* pp. 1153-1157; Rebrov, M. and Khozin, G., *The Moon and Man,* p. 73

13. Petrovich, G.V., *op. cit.*, pp. 122-124

14. Skuridin, G.A., *op. cit.*, pp. 1157-1158

15. *USSR Explores the Moon*, p. 5; Skuridin, G.A., *op. cit.*, pp. 1169-1170; Gatland, K., *op. cit.*, pp. 177-178

16. Results of several scientific instrument readings are documented in Wukelic, G.E., *op. cit.*; Skuridin, G.A., *op. cit.*, p. 1179; Blagonravov, A.A., et al., *op. cit.*, p. 615

17. Petrovich, G.V., *op. cit.*, p. 509; Blagonravov, A.A., et al., *op. cit.*, pp. 614-615

18. Lipskiy, Yu.N., et al., *Atlas of the Reverse Side of the Moon, Part II*, p. 11

19. *Newsweek*, 13 Feb. 1967, p. 25

20. *Spaceflight*, May 1976, p. 188; *New York Times*, 28 Mar. 1969, p. 11

21 *Soviet Space Programs, 1971-1975*, p. 134; *Aviation Week and Space Technology*, 11 Oct. 1971; Turnill, R., *op. cit.*, p. 157

22. *Spaceflight*, May 1976, p. 189; *Soviet Space Programs, 1971-1975*, p. 136; *Space View*, Sep.-Oct. 1973, pp. 22/150-24/152; Gatland, K., *op. cit.*, pp. 211-220; Turnill, R., *op. cit.*, pp. 153-161; *Jane's* estimates the lander assembly to be 907 kg, Wilding-White, T.M., *Jane's Pocket Book of Space Exploration*, pp. 48-51

23. *Space World*, Dec. 1972, pp. 42-45 and Aug. 1972, p. 48 and May 1972, pp. 60-63 and Mar. 1972, p. 42; *Icarus*, June 1973; *Icarus*, Oct. 1972; *Aviation Week and Space Technology*, 3 Jan. 1972, pp. 12-15; *Soviet Space Programs, 1971-1975*, pp. 136-137; *Sky and Telescope*, Feb. 1972. *Russian Report to COSPAR Session, 15th (1981)*, pp. 8-18; *Space Research Performed in USSR in 1972*, pp. 24-37; *Space Research Performed in USSR in 1973*, pp. 22-26. (These sources also include many results of previously mentioned experiments.)

24. *Soviet Space Programs, 1971-1975*, p. 136; *Aviation Week and Space Technology*, 3 Jan. 1972, p. 12; Wilding-White, T.M., *op. cit.*, p. 51

25. *Icarus*, June 1973; *Jane's All the World's Aircraft, 1073-1974; Soviet Space Programs, 1971-1975*, pp. 134-135; Smolders, P., *Soviets in Space*, p. 231; *Spaceflight*, July 1973, pp. 242-245

26. *Sky and Telescope*, Feb. 1972; *Aviation Week and Space Technology*, 3 Jan. 1972, pp. 12-14; *Space Research Performed in USSR in 1972*, p. 26; Gatland, K., *op. cit.*, p. 217; *Aviation Week and Space Technology*, 7 Feb. 1972

27. *Icarus*, June 1973, p. 168

28. *Aviation Week and Space Technology,* 13 Dec. 1971; *Icarus,* June 1973, p. 168

29. *Space Research Performed in USSR in 1972,* p. 28; Gatland, K., *op. cit.,* pp. 217-220; *Astronautics and Aeronautics, 1972,* p. 299; *Icarus,* Oct. 1972, pp. 408-422

30. Sokolov, S.S., *Soviet Automatic Interplanetary Stations Investigate Mars,* pp. 4-5; *Soviet Life,* July 1976

31. *Spaceflight*, May 1976, p. 189

32. *Space Research Conducted in USSR in 1974,* pp. 4-5; *Conquest of Outer Space in the USSR, 1974,* pp. 384-435

33. *Space Research Conducted in USSR in 1974*, p. 5

34. *Space Research Conducted in USSR in 1974,* p. 3; Sokolov, S.S., *op. cit.,* p. 9

35. *Astronautics and Aeronautics, 1973,* p. 267

36. *Soviet Space Programs, 1971-1975,* pp. 138-139; *Space Research Conducted in USSR in 1974,* p. 5. (Initial reports also had Mars 6 carrying Stereo equipment, *Astronautics and Aeronautics, 1973,* p. 242.)

37. Sokolov, S.S., *op. cit.,* pp. 9-10, 16

38. *Aviation Week and Space Technology,* 25 Mar. 1974 and 18 Feb. 1974; *Soviet Space Programs, 1971-1975,* p. 138; *Space Research Conducted in USSR in 1974,* p. 2

39. *Space Research Conducted in USSR in 1974,* pp. 2-3; *Soviet Space Programs, 1971-1975,* p. 139; *Aviation Week and Space Technology,* 25 Mar. 1974

40. *Space Research Conducted in USSR in 1974,* p. 3; *Soviet Space Programs, 1971-1975*, p. 139

41. *Jane's All the World's Aircraft, 1974-1975*; Sokolov, S.S., *op. cit.,* pp. 16, 18-19; *Aviation Week and Space Technology,* 18 Mar. 1974; *Soviet Space Programs, 1971-1975,* p. 139

42. Many of the other scientific results and instruments used are described in the following: *Space Research Conducted in USSR in 1974,* p. 4, 6-21; Sokolov, S.S., *op. cit.,* pp. 22-30; *Soviet Space Program, 1971-1975;* pp. 139-142; *Conquest of Outer Space in the USSR, 1974,* pp. 384-435

43. For just a sample of these predictions see *Soviet Space Programs, 1971-1975*, p. 536; Ponomarev, A.N., *Years of the Space Era*, p. 387; and *Space World*, June 1971, pp. 45-46; *Spaceflight*, June 1973, p. 227.

44. *Spaceflight*, Sep. 1977, pp. 315-316.

Appendix I
SOVIET SPACE LAUNCH VEHICLES

Appendix I

SOVIET SPACE LAUNCH VEHICLES

Since the launch of the simple Sputnik 1 atop a Soviet ICBM in 1957 a wide variety of rockets has been employed for the many diverse missions of the Soviet civilian and military space programs. Since, until 1965, all Soviet spacecraft were launched by military missile derivatives, the USSR had been very reluctant to disclose their configuration or performance statistics. This led to several speculative drawings of these vehicles by Western analysts relying largely on the experience of the United States orbital programs. When the Sputnik booster was finally unveiled in 1967, the futility of this exercise was readily apparent. Soviet rocket designs differed radically from the track followed by Western nations. Details of the older launch vehicles have slowly been made public, although a full photograph of the largest Soviet booster (in use since 1965) has yet to be released (Figure 1).

Due to the sparsity of information in the early years of the Space Age, different classification systems arose to describe the capabilities of these launch vehicles as evidenced by their satellite payloads and military potential. Finally, in 1968 Charles S. Sheldon II published an identification system for Soviet launchers in an article for the 1968-1969 Winter TRW Space Log.[1] This system is now widely recognized and utilized by most Western civilian analysts. Unfortunately, the unknown nature of the later Soviet launch vehicles and differentiation between payload and rocket stage have resulted in minor variations of the Sheldon system by its users. A simplified version will be outlined here.[2]

Sheldon assigned letter designators (A, B, C, etc.) to the different basic launch vehicles while reserving numbers (1, 2, etc.) for the upper orbital stages. Lower case letters could also be added to indicate final stages of specialized purposes, e.g., earth escape capabilities (e), maneuverability (m), or reentry (r).

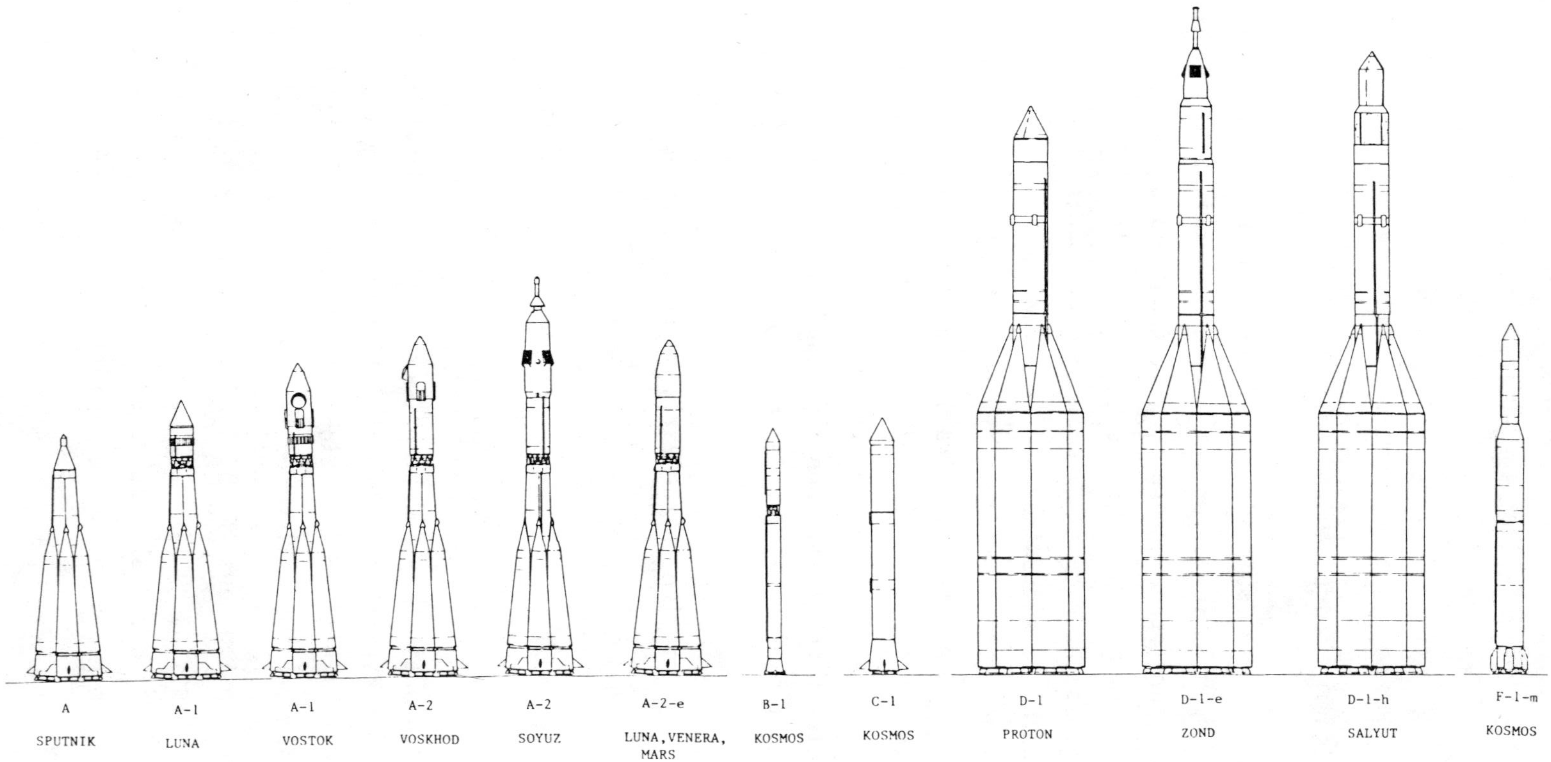

Figure 1 The arsenal of Soviet space launch vehicles (R.F. Gibbons)

It appears possible that a new class of launch vehicle was introduced in 1977 at Plesetsk. In 1977 and 1978 five launches of unknown rockets are thought to have taken place. However, at the time of this writing not enough confirmed data are available to make an exact determination for the purposes of adding a new launch vehicle classification.

A-CLASS VEHICLE

The oldest and still most widely used Soviet space rocket, the A-Class launch vehicles have been successfully used on over 700 orbital missions. These missions have spanned the entire spectrum of earth orbital and deep space objectives, including the first artificial earth satellite, the first lunar, Mars, and Venus robot explorers, and every manned Soviet space flight as well as communications, weather, reconnaissance, and scientific satellites.

The specifications of the A-class vehicle (and subsequent Soviet launchers) can be found in Tables I and II of this Appendix. This basic first generation ICBM was drafted into service in 1957 and 1958 to orbit Sputniks 1 - 3. However, the A launcher (as this configuration was subsequently designated) merely represented a preliminary attempt to investigate the problems and consequences of orbital space flight. Additional upper stages were already under development to increase the payload capability and versatility of the A-class family to include sophisticated unmanned earth satellites, manned spacecraft, and interplanetary explorers. By January 1959 the first A-class upper stage was completed and had successfully propelled Luna 1 on its historic voyage to the moon and beyond.

The direct ascent trajectories followed by Lunas 1-3 to the moon were limited and did not fully utilize the potential lifting ability of the A-class launcher. The technique used to launch all Soviet lunar and planetary probes since Luna 3 has involved the placing of the automated spacecraft in orbit about the earth still attached to a special escape stage. Then after a partial revolution or several revolutions the escape stage fires, hurling the automated probe to its destination. Unfortunately, some developmental problems arose with this final escape stage, resulting in the stranding of many of the early planetary and lunar probes in earth orbit. Here they faced an early death when their "temporary" orbits soon decayed, destroying the spacecraft upon reentry into the earth's atmosphere.

Table IA SOVIET ROCKET ENGINES

Name	Propellants	Configuration	Thrust (tons)	I_{sp} (m/s)	Chamber Pressure (atm)	Expansion Ratio	Launch Vehicle Usage
RD-107	Kerosene/LOX	4 main nozzles 2 vernier nozzles	102	3077	60	150	A-class 1st stage strap-on
RD-108	Kerosene/LOX	4 main nozzles 4 vernier nozzles	96	3087	52	150	A-class core stage
RD-111	Kerosene/LOX	4 main nozzles (gimballed)	166	3107	80	?	1st stage of unkown launcher
RD-119	UDMH/LOX	1 main nozzle 6 vernier nozzles	11	3450	80	1350	B-class 2nd stage
RD-214	Kerosene/Nitric Acid	4 main nozzles	74	2587	45	64	B-class 1st stage
RD-216	UDMH/Ntirc Acid ?	?	100 ?	?	?	?	C-class 1st stage
RD-219	UDMH/Nitric Acid	2 main nozzles	90	2871	75	268	F-class 2nd stage
RD-253	UDMH/N_2O_4 ?	?	250 ?	?	100's ?	?	D-class 1st stage strap-on

LOX: Liquid Oxygen
UDMH: Unsymmetrical Dimethylhydrazine

Table IB SOVIET LUNAR AND PLANETARY PROPULSION SYSTEMS

Spacecraft	Year Introduced	Thrust (kg)	I_{sp} (m/s)	Oxidizer	Fuel	Combustion Chamber Pressure (atm)	Number of Nozzles	Type of Propellant Feed	Number of Engine Firings Permitted
Venera, Mars, Zond (1st generation)	1961	200	2670	Nitric Acid	UDMH	12	1	Pressure	
Venera, Mars (2nd generation)	1971	1000 - 1900	3090			95	1	Pump	Up to 7
Luna (2nd generation)	1963	4640	2720	Nitrous Oxide ?	Amine-based	64	1	Pump	
Luna (3rd generation)									
- Main Landing	1969	750 - 1930	3080		UDMH		1	Pump	Up to 11
- Low Thrust Landing	1969	210-350	2490		UDMH		2	Pressure	1
- Take-off	1970	1920			UDMH	94	1	Pump	

Table II SOVIET SPACE LAUNCHERS

Designator	Launch Facilities	Payload Capability (kg)	First Year in Operation	Major Payloads Carried
A-class vehicles (based on SS-6, Sapwood ICBM)				
A	TT	1500 to earth orbit	1957	Sputniks 1-3
A-1	TT, PL	5000 to earth orbit 400 to moon	1959	Vostok, 1st generation Lunas, Meteor (weather), Kosmos (reconnaissance)
A-2	TT, PL	7500 to earth orbit	1963	Voskhod, Soyuz, Progress, Kosmos (reconnaissance)
A-2-e	TT, PL	7500 to earth orbit 1800 to moon 1200 to Venus, Mars	1961	2nd generation Lunas, 1st generation planetary probes, Molniya (communications), Kosmos (early warning, communications)
B-class vehicles (based on SS-4, Sandal IRBM)				
B-1	PL, KY	450 to earth orbit	1962	Kosmos (weather, military), Interkosmos
C-class vehicles (based on SS-5, Skean LRBM)				
C-1	TT, PL, KY	1100 to earth orbit	1964	Kosmos (science, navigation, communications, satellite killer targets)
D-class vehicles (not related to operational military missile)				
D-1	TT	20,000 to earth orbit	1965	Proton 1-4
D-1-e	TT	20,000 to earth orbit 6000 to moon 5000 to Venus, Mars	1967	Zond 4-8, 3rd generation Lunas, 2nd generation planetary probes, geosynchronous satellites
D-1-h	TT	22,500 to earth orbit	1970	Salyut
F-class vehicle (based on SS-9, Scarp ICBM)				
F-1-m	TT	5000 to earth orbit	1966	Kosmos (ocean surveillance, satellite killer interceptors and targets)
F-1-r	TT	4500 to earth orbit	1967	FOBS (fractional orbital bombardment system)
G-class vehicles (yet to be flown successfully; estimated values only)				
G-1	TT	100,000 - 135,000	?	Space station components
G-1-e	TT	100,000 - 135,000 to earth orbit 45,000 - 50,000 to the moon	?	Manned lunar spacecraft

TT: Tyuratam PL: Plesetsk KY: Kapustin Yar

While the new deep space escape stage was undergoing development in the late 1950's and very early 1960's, a new improved upper stage was also being designed to replace the smaller, less powerful stage used by Luna 1-3 and the Vostok spacecraft. Whereas the older A-1 launcher could orbit a 5000-kg payload, the new A-2 launcher was able to lift 7500 kg into a low earth orbit. This later launch vehicle has been used by all the manned Voskhod and Soyuz flights as well as a multitude of civilian and military unmanned satellites.

After careers of 20, 18, and 16 years for the A-1, A-2-e, and A-2, respectively, these launch vehicles are still the mainstay of the Soviet space program and will probably remain so for many years to come.

B-CLASS VEHICLE

The Soviet Union's second launch vehicle did not make its debut until 1962. The introduction of the B-1 launcher not only signaled a growing flexibility of the Soviet space program, but its first flight inaugurated the Kapustin Yar launch complex and the multi-faceted Kosmos series. The endurance of the A-class launch vehicle for orbiting payloads in the 2000-7000-kg range attests to its efficiency both economically and functionally. However, many scientific or militarily oriented spacecraft weigh much less than 2000 kg. Thus, a smaller and more economical rocket was needed to accommodate the requirements of smaller payloads.

This simple two-stage rocket has been successful in 144 orbital launches during the past 17 years. Payloads of up to 450 kg have been handled effectively. Initially flown only from the launch facilities at Kapustin Yar, the B-1 appeared at Plesetsk a year after this new cosmodrome was opened. Since 1974, all B-1 launches have been exclusively from Plesetsk.

In its early days the B-1 launcher was largely devoted to scientific spacecraft in the Kosmos series. Later, when the international Interkosmos program was begun, the B-1 was responsible for orbiting seven of the first nine of these probes. Although the B-1 supported some military spacecraft from Kapustin Yar, since the move to Plesetsk the duties of this rocket have been almost exclusively of a military nature.

No variation of the B-class vehicle from the original B-1 has been detected. Although not a heavily used rocket (in only four year have there been no more than a dozen flights per year), the B-1 appears still to have found favor in its simplicity and reliability and expense. A recent decrease in its use may signal the departure of this launch vehicle from the Soviet arsenal. If true this would represent the first retirement of any Soviet launcher since the Space Age began.

C-CLASS VEHICLE

The third Soviet space launcher has been designed to fill the void between the small capacity B-class vehicle and the much larger A-class vehicle. About the same height as the B-1, the C-1 launcher has roughly twice the diameter and can lift spacecraft weighing up to 1100 kg into low earth orbit. In accordance with the philosophies of the A- and B-class launchers, the design of the C-1 was based on a previously developed nuclear-tipped military missile, the SS-5 LRBM.

The C-class space rocket is the second most widely used launch vehicle and is the only launch vehicle to have launch facilities at all three Soviet cosmodromes. A wide variety of payloads has been carried by the C-1, including civilian scientific spacecraft (Kosmos and Interkosmos), navigational satellites, electronic intelligence gathering satellites, and on occasion target vehicles for the testing of the Soviet satellite killer system.

Like the B-class launcher, the C-class launcher is a simple two stage rocket and has only been seen in the C-1 configuration. The C-1 is widely known for its frequently demonstrated ability to orbit eight different satellites at one time. Recent usage of the C-1 indicates that this rocket will continue to be employed for quite some time in the orbiting of intermediate size satellites. Sometime in 1979 the C-1 launcher should orbit its 200th payload.

D-CLASS VEHICLE

The 7500-kg capacity of the A-class vehicle was obviously inadequate to fulfill the needs of the larger spacecraft which the Soviet Union would have to manufacture in order to maintain her prestige in space exploration. Although advanced lunar and planetary explorers and geosynchronous satellites launched by the USSR to date are less than 6000 kg, their trajectories and orbital requirements place demands on the launch vehicle far in excess of the capabilities of the A-class launchers. In addition, even small manned space stations or space station modules weigh in the tens of thousands of kilograms--again beyond the capabilities of the A-class rocket.

Therefore, by the early 1960's plans for a new space rocket with a lifting ability three times that of the A-2 (then the rocket with the largest payload capacity) were finalized. In 1965 this rocket made its maiden voyage, placing the 12,200-kg Proton 1 satellite into earth orbit. Since that time, this launch vehicle has been named the Proton launch system by the Soviets and is often referred to simply as the Proton rocket (regardless of its upper-stage configuration).

The D-class vehicle is unique in the present day Soviet space rocket arsenal in that it cannot be directly related to any operational military missile. Although occasional reference has been made linking the D-class vehicle with a Khruschev-era ICBM capable of delivering a 100-megaton "city buster" hydrogen bomb, there is no direct evidence of this, and any such military project was surely abandoned at an early stage.

The D-class launch vehicle has yet to be shown in its entirety even 14 years after its first flight. As a consequence, the designations of different D-class configurations are not held universally owing to different interpretations of rocket construction. As few as two and as many as five different designators are currently used to describe D-class launchers: D, D-1, D-1-e, D-1-h, and D-1-E-e. In Table II of this appendix, only the D-1, D-1-e, and D-1-h designators have been employed to account for all D-class launches to date. In fact, the distinction between D-1 and D-1-h is itself tenuous.

D-class vehicle payloads fall into three broad categories: heavy low-earth orbit spacecraft, geosynchronous satellites, and lunar/planetary explorers. The first category includes the original namesake scientific satellites Protons 1-4, Kosmos 382, and the man-related Salyut space stations. It is here assumed that Protons 1-4 were launched by a D-1 rocket while

Kosmos 382 and Salyuts 1-6 were launched by a slightly modified D-1-h (the "h" standing for "high energy" which is snyonymous with greater lifting power). The D-1-h vehicle should be capable of placing a 20,000 kg payload into a low-earth orbit.

Geosynchronous satellites and lunar/planetary explorers are orbited by the D-1-e variation. In the case of the former payloads, the "escape" stage is employed in the complicated maneuvers required to place a satellite launched from within the USSR into a stationary orbit over the equator. Such a rocket configuration has sent spacecraft weighing 6000 kg to the moon, 5000 kg to Mars and Venus, and 2100 kg to equatorial inclinations.

Facilities to launch D-class vehicles which probably weigh nearly 1,000,000 kg are located only at Tyuratam. By the end of 1978, 55 spacecraft had been successfully launched by this massive rocket with as many as nine more launch failures.

F-CLASS VEHICLE

The fifth and thus far latest confirmed class of Soviet space launcher to successfully fly is the F-series. Although 60 spacecraft have been orbited by F-class vehicles since its introduction in 1966, all payloads fall into three narrow military categories: the satellite interceptor (sic, killer) system (SIS), the fractional orbital bombardment system (FOBS), and the ocean surveillance system (OSS).

The SIS and OSS programs are regarded as employing the F-1-m launcher due to the orbital maneuvers demonstrated by both these systems. However, the design of the upper "m" stage is likely to be slightly different for the SIS and OSS programs as a result of the different type of maneuvers performed. Meanwhile, the short-lived FOBS program utilized a F-1-r launch vehicle since its primary mission was to send nuclear warheads through reentry and back to earth after completing less than one complete revolution about the earth.

The F-1-m appeared in 1966 and has a payload capability of almost 5000 kg. Arriving on the space scene the following year, the F-1-r has the power to place a 4500 kg payload into orbit. Again, the F-class launch vehicle was based on the then recently deployed strategic missile, the SS-9 ICBM.

Today, the FOBS program has long since been active with no confirmed flight of the F-1-r vehicle since 1971. (The proposed SALT II treaty recognizes the existence of this system and calls for its abolishment). Although the SIS is an on-again, off-again project, the F-1-m vehicle continues to find regular use in the Soviet ocean surveillance system.

G-CLASS VEHICLE

Although this vehicle has yet to be successfully test-flown, its existence has been the topic of wide-spread debate for more than a decade and a half, warranting a brief discussion here. As powerful as the D-class launcher is, US analysts were quick to point out that it was incapable of supporting manned lunar missions unless costly, and perhaps difficult, earth orbital assembly was undertaken. Since the world assumed (and all evidence supports) that the Soviets were indeed involved in a race to land men on the moon, some larger launch vehicle must necessarily have been under development for use in the late 1960's.

Affectionately referred to as "Webb's Giant" after the former NASA Administrator James Webb's mention in the mid-1960's of an imminent Soviet Saturn V-class launcher, the G-class launch vehicle is widely reported to have undergone at least three launch attempts--all of which supposedly ended in spectacular failure. A complete review of the reports, predictions, and ramifications of the G-class vehicle is beyond the scope of this short appendix, but its probable general characteristics will be outlined.[3]

The debate conducted in the United States to decide the method of landing and returning the Apollo astronauts during their lunar expeditions was a heated one. Ultimately, the Lunar Orbital Rendezvous (LOR) method won out, but only after the Earth Orbital Rendezvous (EOR) and direct ascent trajectories had been proven beyond the capabilities of a single Saturn V. The same physical laws, of course, applied to the Soviet Union's efforts. Whether the Soviets also chose the LOR technique is not at all clear, but to permit any kind of analysis for the G-class launch vehicle, a performance roughly equal to that of the US Saturn V is assumed.

During the 1969-1972 period when the US Apollo missions were flown to the moon, the total weight of the Apollo/Lunar Module combination ranged from

44.5 to 47 metric tons. This represents nearly the minimum weight for any manned spacecraft launched by a single rocket and designed to land on the moon and return to earth. Therefore, a G-1-e ("e" indicating earth-orbital escape capability) launch vehicle is presumed to have 45,000 - 50,000-kg maximum payload after translunar injection. Similarly, a G-1 type launch vehicle should be capable of sending 100,000 - 135,000 kg into a low-earth orbit.

The reputed history of the G-class vehicle has been inglorious indeed. After repeated delays the first attempted launching is said to have occurred in early June 1969, but ended in disaster when the rocket exploded while still on the ground, demolishing not only the rocket itself, but also the launching complex. The next attempt reportedly came in the summer of 1971. This try, however, proved little better--the rocket exploding in flight while still under power of the first stage. The last G-class launching is said to have come on 24 November 1972 when the rocket was once again destroyed during the initial portion of its flight.

Since 1972, periodic stories have surfaced predicting the imminent launching of a redesigned G-class vehicle. To date such prophecies have yet to come to pass. Apparently, activity around the old G-class launching complex has increased in recent years, giving rise to these stories. One theory is that this site is being refurbished for the announced Soviet space shuttle. It is possible that the first generation space shuttle launches will utilize a derivative of the older G-class vehicle as a disposable booster.

Table III SOVIET LAUNCH VEHICLE ACTIVITY

(Successful Launches per Year by Launch Vehicle Class)

	A	B	C	D	F
1957	2				
1958	1				
1959	3				
1960	3				
1961	6				
1962	13	7			
1963	13	4			
1964	22	7	1		
1965	33	7	6	2	
1966	34	7	0	1	2
1967	36	13	4	2	11
1968	41	16	6	4	7
1969	43	14	6	4	3
1970	43	18	10	4	6
1971	40	12	19	6	6
1972	47	12	13	1	1
1973	53	10	15	7	1
1974	49	6	17	6	3
1975	58	5	18	4	4
1976	54	4	28	5	8
* 1977	54	2	28	4	7
* 1978	59	0	21	5	1
	707	144	192	55	60

* *Three launches in 1977 and two launches in 1978 were launched by unknown rockets. These flights are not included in this table.*

Table IV SOVIET MILITARY LAUNCHERS

Designator	Code-name	Est. Year Operational	Type	Propellants	Range (km)	Comment
SS-1	Scunner	1947	MRBM	liquid	350	
SS-1B	Scud A	1957	MRBM	UDMH/IRFNA	130	
SS-1C	Scud B	1957	MRBM	UDMH/IRFNA	280	
SS-2	Sibling	1948	MRBM	?	?	Very few deployed
SS-3	Shyster	1955	MRBM	Kerosene/O_2	900	
SS-4	Sandal	1959	IRBM	Kerosene/nitric acid	1800	
SS-5	Skean	1961	LRBM	UDMH/nitric acid ?	3500	
SS-6	Sapwood	1957	ICBM	Kerosene/O_2	8500	1st generation ICBM
SS-7	Saddler	1961	ICBM	liquid	10,500	
SS-8	Sasin	1963	ICBM	liquid	11,000	
SS-9	Scarp	1965	ICBM	storable liquid	12,000	First 2nd generation ICBM
SS-10	Scrag	1965	ICBM	Kerosene/O_2	12,000	Not deployed
SS-11	Sego	1966	ICBM	storable liquid	10,500	
SS-12	Scaleboard	1968	MRBM	solid	700-800	
SS-13	Savage	1968	ICBM	solid	8000	
SS-14	Scapegoat	1968	LRBM	solid	4000	Upper two stages of SS-13
SS-15	Scrooge	1969	LRBM	solid	5600	Modified SS-13
SS-16		1977	ICBM	solid	9500	Replacing SS-13
SS-17		1975	ICBM	storable liquid	10,500	Replacing SS-11
SS-18		1976	ICBM	storable liquid	12,000	Replacing SS-9
SS-19		1975	ICBM	storable liquid	10,000	Replacing SS-11
SS-20		1976	LRBM	solid	7500	Lower stages of SS-16
SS-21		1979	Tactical		120	Replaces Frog short range tactical surface to surface
SS-22		1979	MRBM			Replaces SS-12

REFERENCES AND NOTES

1. Mr. Sheldon, then acting Chief of the Science Policy Research Division, Legislative Reference Service of the Library of Congress, is now Chief of the Science Policy Research Division, Congressional Research Service of the Library of Congress and is considered one of the foremost US experts on the Soviet space program.

2. See the following for variations of the Sheldon system:

   TRW Space Log, Winter 1968-1969, pp. 8-19; *Soviet Space Programs, 1971-1975,* pp. 39-64; Turnill, R., *The Observer's Spaceflight Directory,* pp. 336-337; *Spaceflight,* Feb. 1977, pp. 54-60, 80

3. Several very good attempts have been made to examine the "history" of this fascinating launch vehicle. See:

   Clark, P.S., *The Proton and Type-G Launch Vehicles, A Review - Spaceflight,* Dec. 1973, pp. 457-471 and Mar. 1974, pp. 94-104. *Soviet Space Programs, 1971-1975,* pp. 61-64, 218-221, 502-515

   For related reports see:

   Aviation Week and Space Technology, 10 Mar. 1969, 17 Nov. 1969, 6 July 1970, 17 Mar. 1975; *Spaceflight,* July-Aug. 1976, p. 237; Baker, D., *The Rocket,* p. 225

Appendix II
SOVIET SPACE LAUNCH FACILITIES

APPENDIX II

SOVIET SPACE LAUNCH FACILITIES

Presently the Soviet Union utilizes three different launching complexes for orbital spacecraft and deep space probes (Figure 1). Over the years, the strictest secrecy has surrounded these sites as evidenced by the use of misleading geographical names, improper coordinates, and the nighttime arrivals and departures of the US officials during the Apollo-Soyuz Test Project (ASTP). In part this is due to the military as well as civilian activities which take place at these centers. Although other testing bases exist for evaluating military rockets (ICBM's, LRBM's, etc.) and smaller weapon systems, classified military orbital payloads are launched from the same three locations as all civilian spacecraft.

Figure 1 The location of the three Soviet launch facilities

TYURATAM

The launch of Sputnik 1 in October of 1957 provided additional evidence for the location of the launch facilities which had also tested the world's first ICBM two months earlier. The exact location of the launchings could not be pinpointed from Sputnik 1 alone, but its presence could be narrowed down to lying somewhere along the initial ground track of that satellite. Here the Soviet Union attempted to (and still does!) mislead Western specialists by christening the launch complex the Baikonur Cosmosdrome after Gagarin's flight in 1961. Not by coincidence, the railroad town of Baikonur did in fact lie on Sputnik 1's first ground track. However, subsequent launches fixed the launching site some 370 km SW of Baikonur, near the town of Tyuratam in Kazakhstan.

United States' earth survey satellites have been able to place the Baiknour Cosmodrome (or Tyuratam complex as the launching facilities are often interchangeably called) at 45.6° N, 63.4° E. Near the Aral Sea this complex is located in the southern portion of Soviet Central Asia to ease launch vehicle requirements for placing payloads in lower-inclination orbits. Unfortunately, this site is still much further north than comparable US launching facilities, resulting in substantial problems for orbiting such spacecraft as communications and early-warning satellites in geosynchronous orbits. In practice, payloads are initially launched into 50-98° inclinations from Tyuratam.

Tyuratam is roughly similar in nature to the facilities of the Kennedy Space Center/Eastern Test Range at Cape Canaveral, Florida. All manned and lunar/planetary missions have originated there. In addition, a wide range of scientific and military satellites are launched every year at Tyuratam. Tyuratam has launched every type of orbital carrier rocket with the exception of the small (450-kg payload) B-1 launcher and the possible new space launcher introduced in 1977.

Mating of the payload and launcher for at least the A-class and D-class vehicles is done horizontally in assembly buildings at the cosmodrome. In the case of Soyuz (and Vostok and Voskhod before it) this takes place in the MIK (Montazhno Ispytatel'nyy Korpus) some 1.6 km from the launching pad. After all preliminary checkouts have been completed, the entire configuration is transported again horizontally by rail to the launch site. Erection of the rocket and final preparations then ensue.

KAPUSTIN YAR

In March 1962 the broad Kosmos program was inaugurated with the launch of Kosmos 1 by a new space rocket (B-1) from a new space launching facility. Located north and to the west of the Baikonur Cosmodrome, the new launching grounds at Kapustin Yar (48.4° N, 45.8° E) is the smallest of the three Soviet complexes. From a maximum of eight launches in 1968, Kapustin Yar has orbited only 1 or 2 payloads per year throughout most of the seventies. Only the two smallest Soviet space rockets, the B-1 and the C-1 are launched from there. Payload types include Soviet and Interkosmos scientific spacecraft and small military satellites.

Kapustin Yar dates back to pre-Sputnik days as a rocket test center. This site is also used to send up high-altitude geophysical probes and to evaluate space related military systems. Satellites from Kapustin Yar are placed into orbits with inclinations between 48.9° and 50.6°.

PLESETSK

The newest launching facility and now the world's most productive launch site, Plesetsk is located still further north, near the Artic Circle at 62.8° N, 40.1° E. Devoted primarily to orbiting militarily oriented satellites, such as photo-reconnaissance, meteorlogical, navigation, electronic ferret, Plesetsk is also the site from which many of the Molniya and Interkosmos missions originate.

Since its first orbital launching in 1966, Plesetsk had launched 815 space rockets successfully by the end of 1978. After only three years in operation Plesetsk was launching more rockets than Tyuratam and Kapustin Yar combined. During the eight-year period, 1971-1978, launch control personnel averaged placing a space vehicle in orbit once every 4.5 days. From Plesetsk's location close to the Artic Circle spacecraft are placed in earth orbits with high inclinations with respect to the equator: ranging from 62.8° to 82.9°.

Thus far, Plesetsk has demonstrated the capability to launch all operational Soviet space rockets with the exception of the large, civilian D-class launch vehicle. The purported new space launcher seen in 1977 and 1978 has only flown from Plesetsk.

Table I

SOVIET SPACE LAUNCHES TO EARTH ORBIT

	Tyuratam	Kapustin Yar	Plesetsk
1957	2		
1958	1		
1959	3		
1960	3		
1961	6		
1962	13	7	
1963	13	4	
1964	23	7	
1965	41	7	
1966	31	7	6
1967	33	7	26
1968	36	8	30
1969	30	4	37
1970	27	5	48
1971	29	1	53
1972	20	2	52
1973	23	2	61
1974	25	1	55
1975	26	1	62
1976	34	1	64
1977	27	2	69
1978	26	1	61
	472	67	624

Total Launchings: 1163

Appendix III
SOVIET LUNAR AND PLANETARY FLIGHT SUMMARY

Appendix III

SOVIET LUNAR AND PLANETARY FLIGHT SUMMARY

Table 1

SOVIET LUNAR AND PLANETARY EXPLORATIONS

MOON - Luna Series	Successess	Confirmed Failures	Additional Unconfirmed Failures
First-Generation Spacecraft (1959)	3	0	3
Second-Generation Spacecraft (1963-1968)	6	8	3
Third-Generation Spacecraft (1969-Present)	7	5	3
Zond Series			
First-Generation Spacecraft (1965)	1	0	0
Second-Generation Spacecraft (1967-1970)	4	3	2
Sub Total	21	16	11
VENUS - Venera Series			
First-Generation Spacecraft (1961-1972)	5	12	3
Second-Generation Spacecraft (1975-Present)	6*	0	0
Zond Series			
First-Generation Spacecraft (1964)	0	1	0
Sub Total	11	13	3
MARS - Mars Series			
First-Generation Spacecraft (1960-1962)	0	5	0
Second-Generation Spacecraft (1971-Present)	3*	7*	1
Zond Series			
First-Generation Spacecraft (1964-1965)	1#	1	0
Sub Total	4	13	0
Grand Total	36	42	15

* *For the purpose of this table Venera 9, Venera 10, Kosmos 419, Mars 2, and Mars 3 are each counted as two separate spacecraft: one lander & one orbiter.*

This was a systems test only. No data relating to Mars were returned.

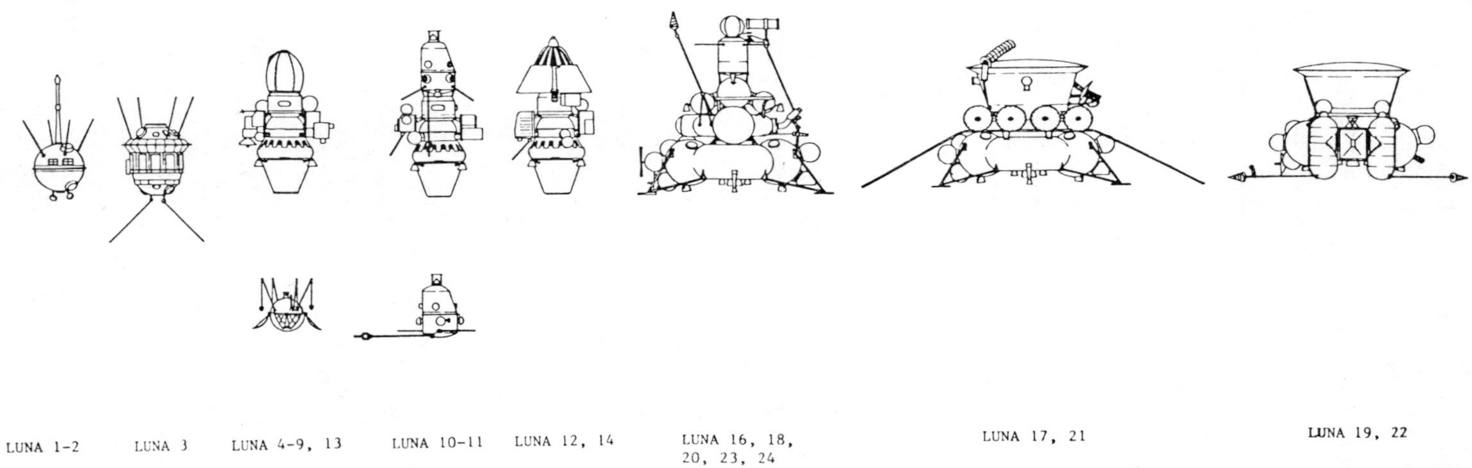

1st generation: Lunas 1-3
2nd generation: Lunas 4-14
3rd generation: Lunas 15-24

Figure 1 The family of Luna explorers (R.F. Gibbons)

Table II

SUMMARY OF SOVIET LUNAR PROBE ATTEMPTS

Designator	Date Launched	Comments
Unannounced	1 May 1958	Possible launch failure
Unannounced	24 Sep 1958	Possible launch failure
Unannounced	26 Nov 1958	Possible launch failure
Luna 1	2 Jan 1959	First lunar flyby
Luna 2	12 Sep 1959	First lunar impact
Luna 3	4 Oct 1959	First photographs of far side
Unannounced	4 Jan 1963	Failed to leave earth orbit
Unannounced	Feb 1963	Possible launch failure
Luna 4	2 Apr 1963	Missed moon by 8500 km
Unannounced	Apr 1964	Possible launch failure
Kosmos 60	12 Mar 1965	Failed to leave earth orbit
Unannounced	Apr 1965	Possible launch failure
Luna 5	9 May 1965	Crashed on moon
Luna 6	8 Jun 1965	Missed moon by 161,000 km
Zond 3	18 Jul 1965	Photographed far side
Luna 7	4 Oct 1965	Crashed on moon
Luna 8	3 Dec 1965	Crashed on moon
Luna 9	31 Jan 1966	First lunar soft landing
Kosmos 111	1 Mar 1966	Failed to leave earth orbit
Luna 10	31 Mar 1966	First lunar satellite
Luna 11	24 Aug 1966	Lunar satellite
Luna 12	22 Oct 1966	Photographic lunar satellite
Luna 13	21 Dec 1966	Lunar soft landing
Kosmos 146	10 Mar 1967	Failed to leave earth orbit
Kosmos 154	8 Apr 1967	Failed to leave earth orbit
Unannounced	22 Nov 1967	Possible launch failure
Zond 4	2 Mar 1968	Unknown mission; probably entered heliocentric orbit
Luna 14	7 Apr 1968	Lunar satellite
Zond 5	14 Sep 1968	Successful lunar flyby and earth return

Table II (Continued)

Designator	Date Launched	Comments
Zond 6	10 Nov 1968	Successful lunar flyby and earth return
Unannounced	8 Jan 1969	Possible launch failure
Unannounced	Apr 1969	Possible launch failure
Unannounced	14 Jun 1969	Possible launch failure
Luna 15	13 Jul 1969	Crashed on moon
Zond 7	8 Aug 1969	Successful lunar flyby and earth return
Kosmos 300	23 Sep 1969	Failed to leave earth orbit
Kosmos 305	22 Oct 1969	Failed to leave earth orbit
Unannounced	19 Feb 1970	Possible launch failure
Luna 16	12 Sep 1970	First automated lunar soil return
Zond 8	20 Oct 1970	Successful lunar flyby and earth return; end of Zond program
Luna 17	10 Nov 1970	First automated lunar rover
Luna 18	2 Sep 1971	Crashed on moon
Luna 19	28 Sep 1971	Photographic lunar satellite
Luna 20	14 Feb 1972	Automated lunar soil return
Luna 21	8 Jan 1973	Automated lunar rover
Luna 22	29 May 1974	Photographic lunar satellite
Luna 23	28 Oct 1974	Failed lunar soil return
Luna 24	9 Aug 1976	Automated lunar soil return

Table III

CONFIRMED SOVIET LUNAR MISSIONS BY YEAR

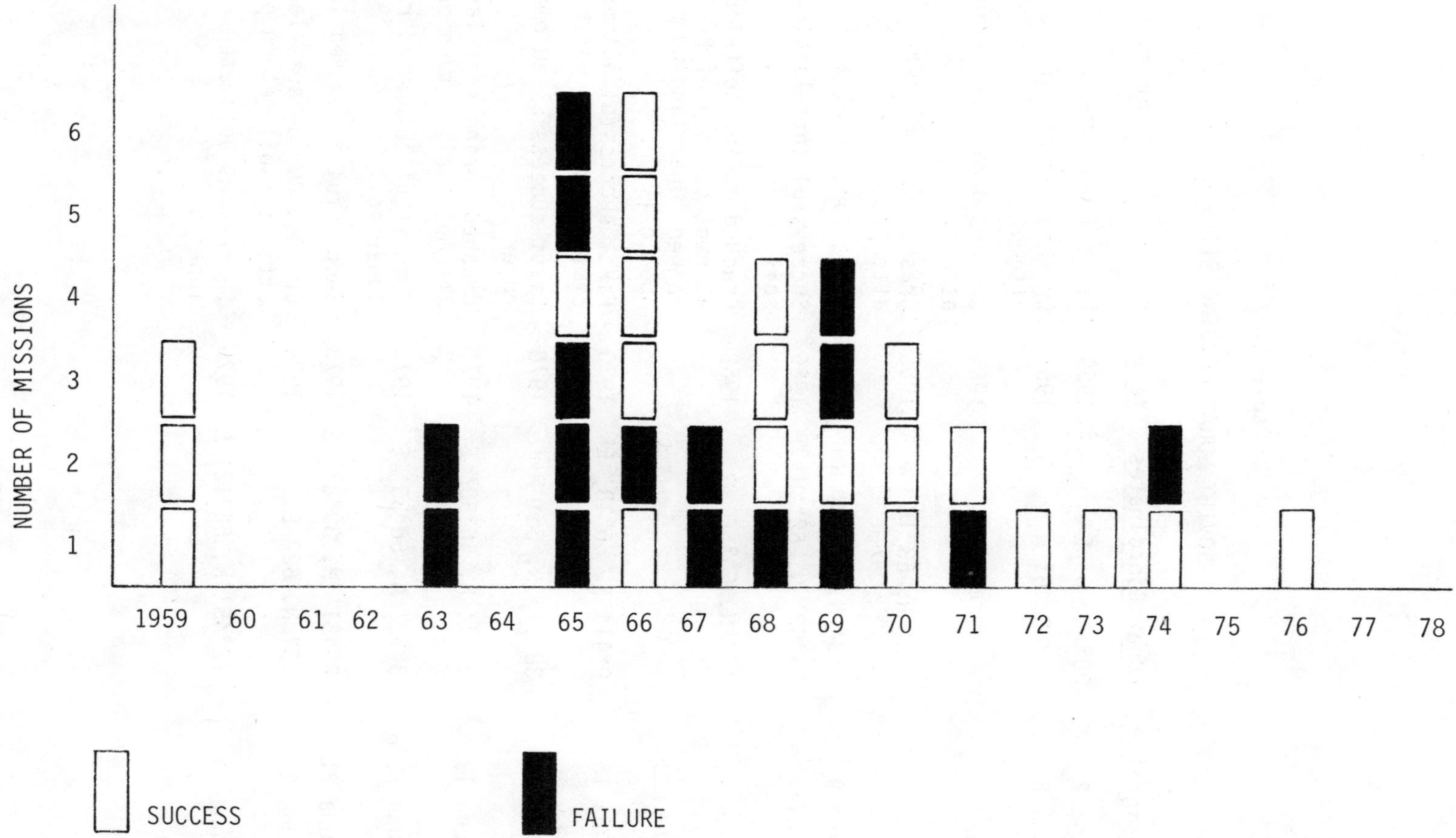

Table IV

SOVIET LUNAR LANDING SITES

Designator	Lunar Coordinates	Year	Comments
Luna 2	30° N, 0° E	1959	First lunar impact
Luna 5	31° S, 8° W	1965	Crashed during soft-landing attempt
Luna 7	9° N, 49° W	1965	Crashed during soft-landing attempt
Luna 8	9°8' N, 63°18' W	1965	Crashed during soft-landing attempt
Luna 9	7°8' N, 64°22' W	1966	First successful soft-landing
Luna 13	18°52' N, 62° W	1966	Successful soft-landing; analyzed soil
Luna 15	17° N, 60° E	1969	Crashed during soft-landing attempt shortly after the first manned lunar landing by the US Apollo 11
Luna 16	0°41' S, 56°18' E	1970	First successful automated return of lunar soil to earth
Luna 17	38°17' N, 35° W	1970	First successful automated lunar rover
Luna 18	3°34' N, 56°30' E	1971	Crashed during soft-landing attempt (soil return probe)
Luna 20	3°32' N, 56°33' E	1972	Successful automated return of lunar soil
Luna 21	25°51' N, 30°27' E	1973	Successful automated lunar rover
Luna 23	≈13° N, 62° E	1974	Crashed during soft-landing attempt (soil return probe)
Luna 24	12°45' N, 62°12' E	1976	Successful automated return of lunar soil

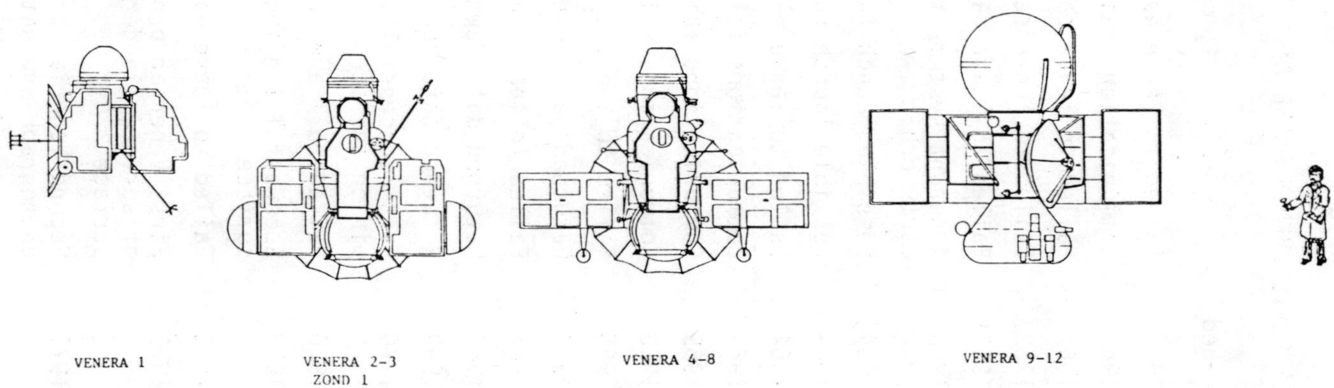

1st generation: Veneras 1-8, Zond 1
2nd generation: Veneras 9-12

Figure 2 The family of Venera explorers (R.F. Gibbons)

Table V

SUMMARY OF SOVIET VENUS PROBE ATTEMPTS

Designator	Date Launched	Comments
Sputnik 7	4 Feb 1961	Failed to leave earth orbit
Venera 1	12 Feb 1961	Communications failed enroute
Unannounced	25 Aug 1962	Failed to leave earth orbit
Unannounced	1 Sep 1962	Failed to leave earth orbit
Unannounced	12 Sep 1962	Failed to leave earth orbit
Kosmos 21	11 Nov 1963	Apparent systems test; failed to leave earth orbit
Unannounced	27 Feb 1964	Possible launch failure
Unannounced	4 Mar 1964	Possible launch failure
Kosmos 27	27 Mar 1964	Failed to leave earth orbit
Zond 1	2 Apr 1964	Communications failed enroute
Venera 2	12 Nov 1965	Communications failed enroute
Venera 3	16 Nov 1965	Communications failed enroute
Kosmos 96	23 Nov 1965	Failed to leave earth orbit
Venera 4	12 Jun 1967	Returned data down to 27 km
Kosmos 167	17 Jun 1967	Failed to leave earth orbit
Venera 5	5 Jan 1969	Returned data down to 24-26 km
Venera 6	10 Jan 1969	Returned data down to 10-12 km
Venera 7	17 Aug 1970	First successful landing; survived 23 min on surface
Kosmos 359	22 Aug 1970	Failed to leave earth orbit
Venera 8	27 Mar 1972	Successful landing; survived 50 min on surface
Kosmos 482	31 Mar 1972	Failed to leave earth orbit
Venera 9	8 Jun 1975	First Venerian orbiter; first photograph of surface; lander survived 53 min on surface
Venera 10	14 Jun 1975	Second Venerian orbiter; lander returned photograph and data for 65 min
Venera 11	9 Sep 1978	Successful landing; survived 95 min on surface
Venera 12	14 Sep 1978	Successful landing; survived 110 min on surface

Table VI CONFIRMED SOVIET VENUS MISSIONS BY YEAR

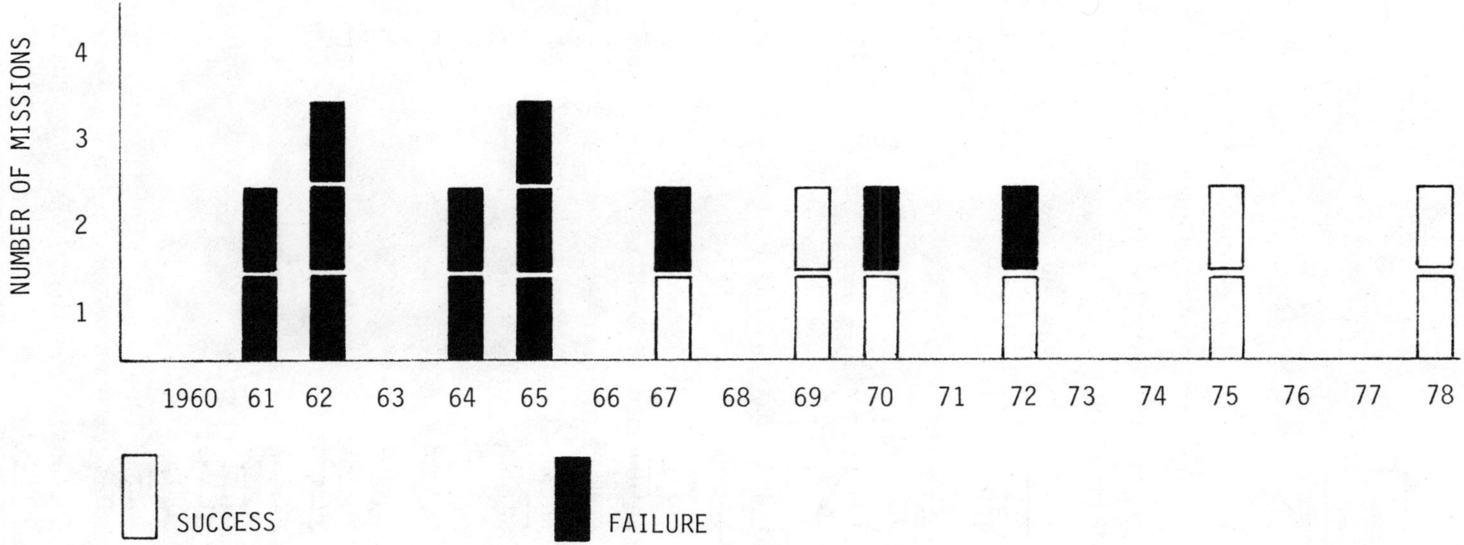

Table VII IDENTIFIED SOVIET VENUS LANDING SITES

Designator	Venus Coordinates (Lat, Long)	Year	Comments
Venera 9	31°42', 290°50'	1975	Survived 53 minutes
Venera 10	16°02', 291°	1975	Survived 65 minutes
Venera 11	-13°, 300°	1978	Survived 95 minutes
Venera 12	-7°, 294°	1978	Survived 110 minutes

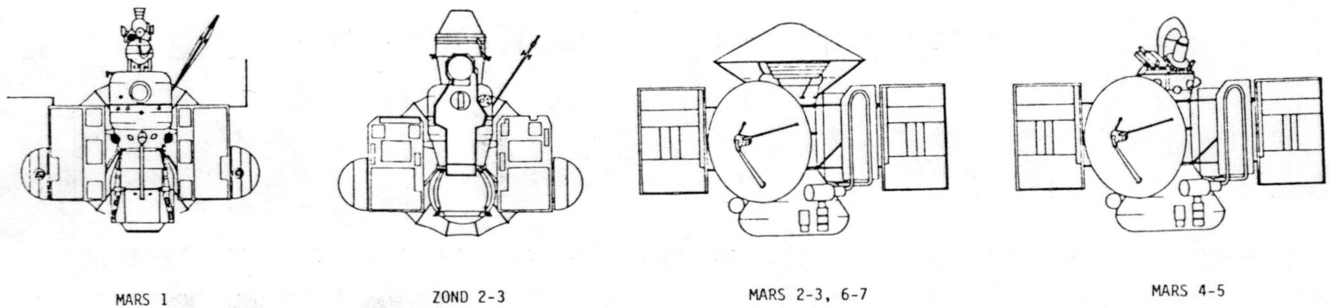

1st generation: Mars 1, Zond 2-3
2nd generation: Mars 2-7

Figure 3   The family of Mars explorers   (R.F. Gibbons)

Table VIII

SUMMARY OF SOVIET MARS PROBE ATTEMPTS

Designator	Date Launched	Comments
Unannounced	10 Oct 1960	Launch failure
Unannounced	14 Oct 1960	Launch failure
Unannounced	24 Oct 1962	Failed to leave earth orbit
Mars 1	1 Nov 1962	Communications failed enroute
Unannounced	4 Nov 1962	Failed to leave earth orbit
Zond 2	30 Nov 1964	Communications failed enroute
Zond 3	18 Jul 1965	Systems test; communications failed 150 million km from earth
Unannounced	27 Mar 1969	Possible launch failure
Kosmos 419	10 May 1971	Failed to leave earth orbit
Mars 2	19 May 1971	Successful Mars orbiter; lander crashed
Mars 3	28 May 1971	Successful Mars orbiter; lander communications failed immediately after landing
Mars 4	21 Jul 1973	Orbiter failed to enter Martian orbit
Mars 5	25 Jul 1973	Successful Mars orbiter
Mars 6	5 Aug 1973	Communications failed immediately prior to landing
Mars 7	9 Aug 1973	Lander missed entire planet

Table IX CONFIRMED SOVIET MARS MISSIONS BY YEAR

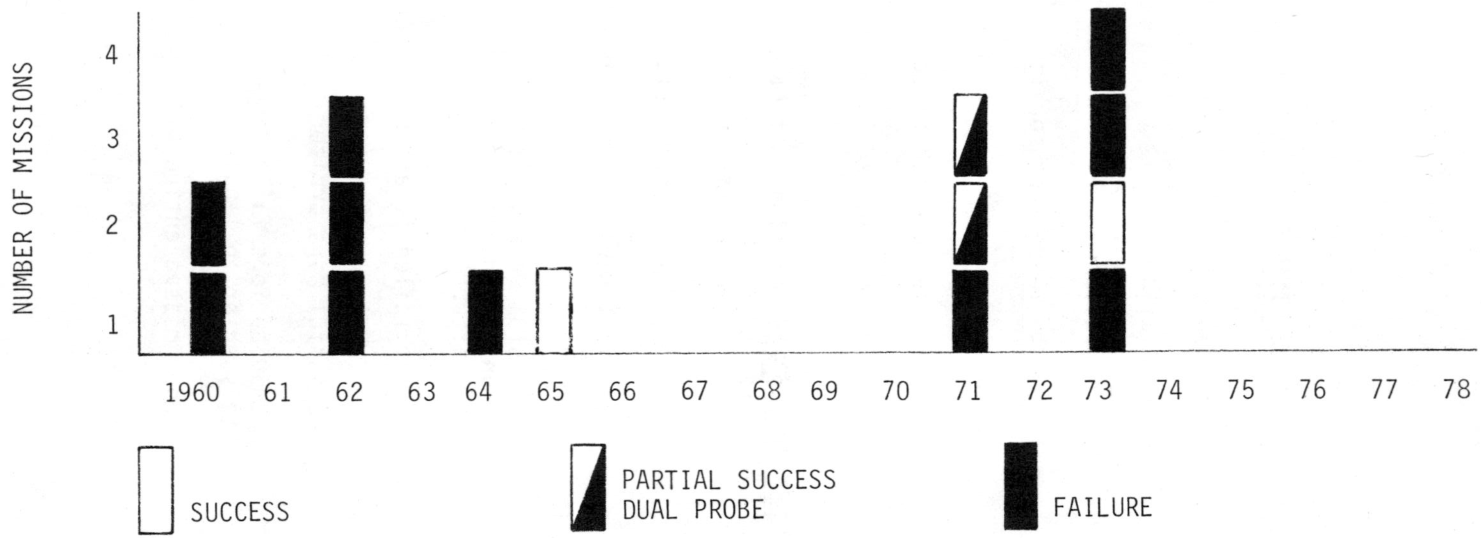

Table X SOVIET MARS LANDING SITES

Designator	Martian Coordinates (Lat. Long)	Year	Comments
Mars 2	45° S, 302° W	1971	Crashed during landing attempt
Mars 3	45° S, 160° W	1971	Ceased transmissions 90 seconds after landing
Mars 6	24° S, 25° W	1974	Ceased transmissions immediately prior to landing

BIBLIOGRAPHY

BIBLIOGRAPHY

The following bibliography contains those books and periodicals which contain significant data concerning the Soviet manned space flight and unmanned lunar and planetary programs. While listing many volumes, this list is by no means exhaustive, but rather is designed to provide a firm base for any further study into these programs. Many of the titles have been translated by the US National Technical Information Service or some other department of the US government or government contractor. The number in parentheses following such works is the NTIS order number by which these books can be purchased. Both paper and microfiche copies are as a rule available.

BOOKS, BOOKLETS, ETC.

Aleksandrov, S.G. and Federov, R.Ye., *Soviet Satellites and Space Ships,* 2nd ed., Moscow, 1961, (N63-18430)

Alekseyev, V. and Lebedev, L., *For Lunar Rock,* Mashinostroyeniye Press, Moscow, 1972, (N73-30801)

Alekseyev, V. and Minchin, S., *Venus Reveals Secrets,* 2nd ed., Mashinostroyeniye Press, Moscow, 1975, (N76-14998)

Andreyanov, V.V., et al., *Automatic Planetary Stations,* Nauka Press, Moscow, 1973, (N74-29290)

Apollo-Soyuz Mission Evaluation Report, NASA Lyndon B. Johnson Space Center, Houston, Texas, 1975. (N76-30242)

Apollo-Soyuz Test Project, Academy of Sciences (USSR), Moscow, 1975, (N75-25985)

Astronautics and Aeronautics, (year), Chronology of Science Technology, and Policy, NASA

Year 1963 - NASA SP-4004
Year 1964 - NASA SP-4005
Year 1965 - NASA SP-4006
Year 1966 - NASA SP-4007
Year 1967 - NASA SP-4008
Year 1968 - NASA SP-4010
Year 1969 - NASA SP-4014
Year 1970 - NASA SP-4015
Year 1971 - NASA SP-4016
Year 1972 - NASA SP-4017
Year 1973 - NASA SP-4018
Year 1974 - NASA SP-4019

Astronauts and Cosmonauts Biographical and Statistical Data, Committee on Science and Technology, US House of Representatives, 1977

An Atlas of the Moon's Far Side, Interscience Publishers, Sky Publishing Company, Cambridge Mass, 1961

Baker, David, *The Rocket, The History and Development of Rocket and Missile Technology,* Crown Publishers, Inc., New York, 1978

Beard, Robert, *Soviet Cosmonautics, 1957-1967,* privately published, 1970

Beard, Robert, *Soviet Space Failures: A Re-Appraisal,* privately published, 1970

Belyakov, I.T. and Dorisov, YU.D., *Technology in Space,* Mashinostroyeniye Press, Moscow, 1974, (N75-25982)

Blagonravov, A.A., Chief Editor, *USSR Achievements in Space Research (First Decade in Space, 1957-1967),* Nauka Publishing House, Moscow, 1968, (JPRS-47311)

Blaine, J.C.D., *The End of an Era in Space Exploration,* Volume 42, Science and Technology Series, American Astronautical Society, 1976

Bono, Phillip and Gatland, Kenneth, *Frontiers of Space,* rev. ed., Macmillan Publishing Company, Inc., New York, 1976

Borisenko, I.G., *In Outer Space,* Mashinostroyeniye Press, Moscow, 1974, (N75-23625)

Borisenko, I.G., *Space Launches and Finishes,* Znaniye Press, Moscow, 1975, (N77-11076)

Bumshtein, S.I., *Salyut Space Station In Orbit,* Mashinostroyeniye Press, Moscow, 1973, (JPRS-62201)

Bychkov, V.N., Nazarov, G.A. and Prishchepa, V.I., *Liquid Rocket Space Engines,* "Kosmonavtika, Astronomiya", No. 9, 1976, (N77-13144)

Caidin, Martin, *Red Star in Space,* Crowell-Collier Press, 1963

Caidin, Martin, *Rendezvous in Space,* E.P. Dutton and Company, Inc., 1962

Clark, Phillip S., *The Proton and Type-G Launch Vehicles,* privately published, January 1977

Supplement 1, June 1977
Supplement 2, January 1978

Clark, Phillip S., *A Review of the Soviet Union's Manned Lunar Programme,* privately published, January 1977

Clark, Phillip S., *Twenty Years of Space Flight, Volumes 1 and 2,* Antares, Volume 2 Special Issue 3, The Bradford Astronomical Society, 1978

Cox, Donald W., *The Space Race,* Chilton Books, Chilton Company, publishers, 1963

Daniloff, Nicholas, *The Kremlin and the Cosmos,* Alfred A. Knopf, New York, 1972

Dimitriyev, A.Yu, et al., *From Spaceships to Orbiting Stations,* Mashinostroyeniye Press, Moscow, 1971, (N74-15535)

Feoktistov, K.P., *The Salyut 4 Space Laboratory,* Vestinik Akademii Nauk USSR, No. 4, April 1976, (N77-10118)

The First Man in Space, Crosscurrents Press, Inc., New York, 1961

First Panoramic Views of the Lunar Surface, Nauka Publishing House, Moscow, 1966, (N66-38486)

Froehlich, Walter, *Apollo-Soyuz,* NASA EP-109, 1976

Gagarin, Yuri, *Road to the Stars,* Foreign Languages Publishing House, Moscow

Gatland, Kenneth, *Astronautics in the Sixties,* John Wiley and Sons, Inc., 1962

Gatland, Kenneth, *Manned Spacecraft,* Macmillan Publishing Company, Inc., New York, 1976

Gatland, Kenneth, *Missiles and Rockets,* Macmillan Publishing Company, Inc., New York, 1975

Gatland, Kenneth, *Robot Explorers,* Macmillan Company Blandford Press Ltd., London, 1972

Gatland, Kenneth, *Spacecraft and Boosters,* Iliffe Books, Ltd., London, 1964

Gazenko, O.G. and Bjurstedt, H.A., eds. *Man in Space,* Moscow, 1974 (available through Univelt, Inc., San Diego, California)

Gherman Titov, The first Man to Spend a Day in Space, Crosscurrents Press, New York, 1962

Gibney, Frank,and Feldman George J., *The Reluctant Space-Farers,* New American Library, New York, 1966

Glushko, V.P., *Rocket Engines GDL-OKB,* Novosti Press Agency, Moscow,1975

Gurney, Gene and Gurney, Clare, *Cosmonauts in Orbit,* F. Watta, New York, 1972

Hall, Al, ed., *Petersen's Book of Man in Space, Volumes 1-5,* Petersen Publishing Company, Los Angeles, California, 1974

Hall, R. Cargill, ed., *Essays on the History of Rocketry and Astronauts: Proceedings of the Third through the Sixth History Symposia of the International Academy of Astronautics*
Volume 1 (N77-33030 thru N77-33047)
Volume 2 (N77-33048 thru N77-33070)

Harvey, Dodd L. and Ciccoritti, Linda C., *U.S. - Soviet Cooperation in Space,* Monographs in International Affairs, Center for Advanced International Studies, University of Miami, 1974

James, Peter N., *Soviet Conquest from Space,* Arlington House Publishers, New Rochelle, New York, 1974

Jane's All the World's Aircraft, Jane's Yearbooks, London, annually

Kamanin, N. and Rebrov, M., *An Experimental Space Station in Orbit, 14-18 January 1969,* Izd-vo TsK VLKSM Molodaya Gvardiya, 1969. (AD-718284)

Khodarev, Y.K., et al., *Equipment for Space Research,* NASA TT-F-785. (N75-10988 thru N75-11016)

Krasnov, N.F., ed., *Rocket Aerodynamics,* Vysshaya Shkola Press, Moscow, 1968, (N71-31467)

Korovkin, A.S., *Spacecraft Control Systems,* Military Press, Moscow, 1972. (N73-25903)

Koval, A.D. and Uspenskiy, G.P., *Space is for Man,* 2nd ed., Mashinostrostroyeniye Press, Moscow, 1974. (N77-11075)

Lebedev, V. and Gagarin, Yuri, *Survival in Space,* Bantom Books, Inc., New York, 1969

Legendary Vostok, Novosti Press Agency Publishing House, Moscow

Leonov, Alexi and Lebedev, Vladimir, *Space and Time Perception by the Cosmonaut,* Mir Publishers, Moscow, 1971

Lewis, Richard S., *From Vineland to Mars, A Thousand Years of Exploration,* Quadrangle/New York Times Book Company, New York, 1976

Ley, Willy, *Events in Space,* Popular Library, New York, 1969

Lipskiy, Yu.N., *Atlas of the Reverse Side of the Moon, Part 2,* Nauka Press, Moscow, 1967, (N69-24206)

Mallan, Lloyd, *Russia's Space Hoax,* Science and Mechanics Publishing Company, New York, 1966

Mandrovsky, Boris, *Soyuz 1. Facts and Speculations,* Aerospace Technology Division, Library of Congress, 1967. (N67-30633)

McGraw-Hill Encyclopedia of Space, McGraw-Hill Book Company, New York, 1968

Moore, Patrick, *Space*, The Natural History Press, 1969

NASA Authorization for Fiscal Year (yearly), Hearings Before the Subcomittee on Science, Technology, and Space of the Committee on Commerce, Science, and Transportation, US Senate; usually contain several references to the Soviet space program

Parry, Albert, *Russia's Rockets and Missiles*, Doubleday and Company, Inc., New York, 1960

Penkovsky, Oleg, *Penkovsky Papers*, Doubleday and Company, Inc., New York, 1965

Peterson, R.W., ed., *Space: From Gemini to the Moon and Beyond*, Facts On File, Inc., New York, 1972

Petrov, G.E., ed., *Conquest of Outer Space in the USSR, 1967-1970*, Amerind Publishing Company Pvt. Ltd., New Delhi, India, 1973

Petrov, V., *Artificial Satellites of the Earth*, Hindustan Publishing Corporation, New Delhi, India, 1960

Petrovich, G.V., ed., *The Soviet Encyclopedia of Space Flight*, Mir Publishers, Moscow, 1969

Ponomarev, A.N., *The Years of the Space Era*, Military Press, Moscow, 1974, (N75-21309)

Rebrov, M. and Khozin, G., *The Moon and Man*, Peace Publishers, Moscow

Riabchikov, Evgeny, *Russians in Space*, Doubleday and Company, Inc., 1971 (Prepared by Novosti Press Agency Publishing House, Moscow)

Romanov, A., *Spacecraft Designer*, Novosti Press Agency Publishing House, Moscow, 1976

Russian Report to the Fifteenth COSPAR Session, Nauka Publishing House, Moscow, 1972, (JPRS-56527)

Sagan, Carl, *The Cosmic Connection*, Dell Publishing Company, Inc., 1975

Sagdeyev, R.Z., ed., *The Conquest of Outer Space in the USSR in 1974*, Nauka Press, Moscow, 1975, (N77-17118)

Salyut Orbital Stations, Novosti Press Agency Publishing House, Moscow, 1975

Schauer, William H., *Politics of Space*, Holmes and Meier, publishers, New York, 1976

Sharpe, Mitchell R., *"It is I, Seagull": Valentina Tereshkova, First Woman in Space,* Thomas Y. Crowell Company, New York, 1975

Sharpe, Mitchell R., *Living in Space, the Astronaut and His Environment,* Doubleday and Company, Inc., New York, 1969

Sheldon, Charles S. II, *United States and Soviet Progress in Space: Summary Data Through 1975 and a Forward Look,* Congressional Research Service, Library of Congress, 1976

Shelton, William, *Soviet Space Exploration: The First Decade,* Washington Square Press, New York, 1968

Shtern, M.I., ed., *Investigations of the Upper Atmosphere and Outer Space Conducted in 1970 in the USSR (14th COSPAR),* Nauka Press, Moscow, 1971, (N72-17970)

Shtern, M.I., ed., *Space Research Conducted in the USSR in 1973 (17th COSPAR),* Nauka Publishing House, Moscow, 1974, (JPRS-62861)

Shternfeld, Ari, *Soviet Space Science (Iskustevenniye Sputniki in the USSR),* Basic Books, Inc., New York, 1959

Sisakyana, N., ed., *Second Group Flight and Certain Results of Flights of Soviet Cosmonauts on the Vostok Ships,* Nauka Publishing House, Moscow, 1965, (AD-619384)

Skuridin, G.A., et al., *Entrance of Mankind into Space (15th Anniversary of the First Manned Flight into Space),* Znaniye Press, Moscow, 1976, (N77-12082)

Skuridin, G.A., ed., *Mastery of Outer Space in the USSR, 1957-1967,* Nauka Press, Moscow, 1971, (N75-29144)

Smolders, Peter L., *Soviets in Space,* Taplinger Publishing Company, Inc., New York, 1974

Sobel, Lester A., *Space: From Sputnik to Gemini,* Facts On File, Inc., New York, 1965

Sokolov, S.S., *Soviet Automatic Interplanetary Stations Investigate Mars,* Vestnik Akademii Nauk USSR, No. 10, (N75-17272)

Solov'yev, Ts. V. and Tarasov, Ye.V., *Interplanetary Flight Prediction,* Mashinostroyeniye Press, Moscow, 1973, (N74-28312)

"The Soviet Interplanetary Station - Venus 4", Pravda, 22 October 1967, (N70-26751)

Soviet Man in Space, Foreign Languages Publishing House, Moscow; *Soviet Moon Rockets,* rev. ed., Soviet Booklet No. 62, Soviet Booklets, Farleigh Press Ltd., London, 1960

Soviet Space Programs, Committee on Aeronautical and Space Science, US Senate, 1962

Soviet Space Programs, 1962-1965, Committee on Aeronautical and Space Sciences, US Senate, 1966

Soviet Space Programs, 1966-1970, Committee on Aeronautical and Space Sciences, US Senate, 1971

Soviet Space Programs, 1971-1975, Committee on Aeronautical and Space Sciences, US Senate, 1976

Soviet Space Research, Novosti Press Agency Publishing House, Moscow, 1975

Soviet Writings on Earth Satellites and Space Travel, The Citadel Press, New York, 1958

Soyuz and Apollo (in Russian), Moscow, 1976

Stockholm International Peace Research Institute (SIPRI), *Outer Space - Battlefield of the Future?,* Crane, Russak and Company, Inc., New York, 1978

Stoiko, Michael, *Soviet Rocketry: Past, Present, and Future,* Holt, Rinehart, and Winston, New York, 1970

Sykes, Margaret W., *Selected Bibliography on Manned Orbital Space Stations,* Space Sciences Laboratory, General Electric Missile and Space Division, 1963, (N64-21677)

Table of Earth Satellites, 1957 - , Royal Aircraft Establishment, Farnborough Hants, England, continually updated

Table of Space Vehicles, 1958 - , Royal Aircraft Establishment, Farnborough Hants, England, continually updated

Ten Years of Space Exploration, Novosti Press Agency Publishing House, Moscow

Thomas, John R. and Kruse-Vaucienne, Ursula M., eds., *Soviet Science Technology: Domestic and Foreign Prospectives,* published for the National Science Foundation by George Washington University, 1977, (PB-276968)

Titov, Gherman, *I am Eagle,* The Bobbs-Merrill Company, Inc., New York, 1962

Titov, Gherman, *700,000 km Through Space,* Foreign Languages Publishing House, Moscow

TRW Space Log, TRW Defense and Space Systems Group, TRW Inc., Redondo Beach, California, currently published annually

Tsiolkovskiy, K.E., *Study of Outer Space by Reaction Devices,* Mashinostroyeniye Press, Moscow, 1967, (N75-29141)

Turnill, Reginald, *The Observer's Book of Manned Spaceflight,* Frederick Warne and Company Ltd., London, 1975

Turnill, Reginald, *The Observer's Book of Unmanned Spaceflight,* Frederick Warne and Company Ltd., London, 1974

Turnill, Reginald, *The Observer's Spaceflight Directory* (a revised and combined edition of the previous two volumes), Frederick Warne and Company Ltd., London, 1978

Umanskiy, S.P., *Man in Space Orbit,* Mashinostroyeniye Press, Moscow, 1975, (N75-10688)

USSR Explores the Moon, Novosti Press Agency Publishing House, Moscow

Vassiliev, M., *Sputnik into Space,* The Dial Press, New York, 1958

Vinogradov, A.P., ed., *Lunokhod 1: Mobile Lunar Laboratory,* (N72-12168)

Vladimirov, Leonid, *The Russian Space Bluff,* The Dial Press, New York, 1973

von Braun, Wernher and Ordway, Frederick III, *History of Rocketry and Space Travel,* 3rd ed., Thomas Y. Crowell, 1975

Voskhod 1 and Voskhod 2 Flights, Surveys of Soviet-Bloc Scientific and Technical Literature, Aerospace Technology Division, Library of Congress, (N65-32217)

Wilding-White, T.M., *Jane's Pocket Book of Space Exploration,* Collier Books, New York, 1976

Wukelic, G.E., *Handbook of Soviet Space-Science Research,* Gordon and Breach Science Publishers, New York, 1968

Yefremov, Yu.I., *The Panoramic Views of the Lunar Surface, Volume 2,* Nauka Press, Moscow, 1969, (N71-19242)

Yefremov, Yu.I., *Space Research Conducted in the USSR in 1974 (18th COSPAR),* Nauka Publishing House, Moscow, 1975, (JPRS-65778)

Yefremov, Yu.I., *Space Research Performed in USSR in 1972 (16th COSPAR),* Moscow, 1973, (JPRS-59778)

Zaehringer, Alfred J., *Soviet Space Technology,* Harper and Brothers Publishers, New York, 1961

Zhdanov, G. and Tindo, I., *Space Laboratories,* Foreign Languages Publishing House, Moscow.

PERIODICALS

Individual articles from periodicals have been indicated in the footnotes at the end of each section. Therefore, the following is a list of those periodicals which frequently or occasionally publish articles pertaining to the space programs of the Soviet Union. The country of origin of each periodical is indicated in parentheses.

Astronautics and Aeronautics (USA)

Astronomicheskii Zhurnal (USSR) (available in USA as *Soviet Astronomy*)

Aviation Week and Space Technology (USA)

Aviatsiya i Kosmonavtika (USSR)

Flight International (England)

Flug Revue (West Germany)

Icarus (USA)

Interavia (Switzerland)

Izvestiya (USSR)

Journal of the British Astronomical Association (England)

Journal of the British Interplanetary Society (England)

Journal of Spacecraft and Rockets (USA)

Kosmicheskie Issledovaniya (USSR) (available in USA as *Cosmic Research*)

L-5 News (USA)

Nature (England)

New Scientist (England)

Pravda (USSR)

Science (USA)

Science News (USA)

Sky and Telescope (USA)

Soviet Aeronautics (USSR) (available in USA in English edition)

Soviet Aerospace (USA)

Soviet Physics - Doklady (Translation of Physics Section of the Proceedings of the Academy of Sciences of the USSR) (USA)

Soviet Life (USA)

Space/Aeronautics (USA)

Spaceflight (England)

Space Technology International (USA)

Spaceview (Holland)

Space World (USA)

Tass (USSR)

Translations on USSR Science and Technology, Physical Sciences and Technology (Translations of Soviet scientific journals by the Joint Publications Research Service, available from NTIS)

BIBLIOGRAPHIES

Whereas many of the volumes referenced above contain excellent bibliographies, the following four bibliographies are worthy of special note due to their comprehensive nature for selected time periods:

Beard, Robert, *Soviet Cosmonautics, 1957-1969, A Bibliography of Articles Published in British Periodicals and of British and Foreign Books,* privately published

Kendon, Anthony, "A Guide to the Study of the Soviet Space Programme", *Spaceflight,* May 1975, pp. 175-179

TRW Space Log, Winter 1968-1969, TRW Defense and Space Systems Group, Redondo Beach, California, pp. 24-29

TRW Space Log, 1974, TRW Defense and Space Systems Group, Redondo Beach, California, pp. 20-35.

INDEX

INDEX